For Marguerite

on the occas[ion]

AF836074

Threatened Plants of New Zealand

IOS visit to Mt Hutt

6.2.1990

Margaret Jo Bulfin

Tertiary limestone at Castle Hill Reserve, Canterbury. Three species at risk occur within the reserve and another three in the immediate vicinity. (D. Given)

DSIR FIELD GUIDE

Threatened Plants of New Zealand

Catherine M. Wilson and David R. Given

Botany Division,
Department of Scientific and Industrial Research.

DSIR Publishing
Wellington 1989

Published by: DSIR Publishing, P.O. Box 9741, Wellington, New Zealand
Edited by: Quentin Ruscoe, DSIR Publishing, Wellington
Designed by: Denis Gourley, DSIR Publishing, Wellington
Typeset by: Comset, Wellington
Printed by: Kings Time Printing Press Ltd, Hong Kong

© Crown Copyright 1989

Cataloguing - in - Publication

WILSON, Catherine M.
Threatened plants of New Zealand/Catherine M. Wilson and David R. Given. –
Wellington: DSIR Publishing, 1989.
(DSIR information series, ISSN 0077-9636; no. 166) (DSIR field guide)
ISBN 0-477-02562-5

I. Title. II. Given, David R. III. New Zealand.
Department of Scientific and Industrial Research. IV. Series. V. Series.

UDC 502.753(931)

Cover photo: Cryptostylis subulata, *duckbill orchid, tongue orchid.*
(E. Cameron)

Foreword

Scientific knowledge grows by adding together small, often disparate, pieces of information. Sometimes, and certainly more rarely, knowledge leaps forward. Both elements are seen in the science of plant conservation - small units of taxonomic information, scattered distributional data, and so on, gave us the first lists of threatened species.

With the addition of breeding biology references, horticultural knowledge, etc., a second generation of more in-depth publications appeared, dealing with threatened species.

At all levels of plant conservation literature, New Zealand has been at the forefront of the science, and it is no coincidence that the name of David Given is to be found as a major author in this advance.

It was, therefore, a real pleasure for me to be asked to write the Foreword for this, the third level or generation of plant conservation works. Each layer of information, by the fact that it is published, excites more interest and attention and, even more importantly, stimulates more detailed knowledge to be identified and presented to aid the next step. The development of a conservation database, which allows the development of appropriate action plans, rational and workable management policies which can be incorporated into development plans, alongside urban and agricultural development plans, is today an integral part of the planning process.

It is at our own peril, both for the quality of life and in ensuring the as yet unknown potential of our "Green Glue", to ignore the plant conservationists' advice.

The authors have reviewed the threats, and carefully listed the species we need to care for and the actions already taken or necessary to ensure that the richness of New Zealand's flora will be there for future generations. It is a resource to be proud of, and I congratulate the authors in capturing the essence of this important topic in one of those leaps forward essential to make all New Zealand citizens aware of the problem and show they can all participate in ensuring tomorrow will see the flowering of today's plant treasures!

G. L. Lucas, O.B.E., B.Sc, F.L.S., F.R.G.S.

Chairman, Species Survival Commission, IUCN (1981-1989)
Keeper, The Herbarium, Royal Botanic Gardens, Kew

Acknowledgments

The compilation of this book was made possible by a generous grant from the New Zealand Lottery Board and the invaluable assistance of numerous individuals.

This field guide synthesises information from many people and therefore, in a sense, it has many authors. We are deeply grateful to Ian Atkinson, the late Jack Bedford, Mark Bellingham, John Braggins, Patrick Brownsey, Margaret Bulfin, Ewen Cameron, Bruce Clarkson, Nigel Clunie, Lindsay Daniel, John Dawson, Peter de Lange, Tony Druce, Chris Ecroyd, Elizabeth Edgar, Alan Esler, Phil Garnock-Jones, Marc Heginbotham, Peter Johns, Peter Johnson, Graeme Loh, David Lloyd, Bryony Macmillan, Alan Mark, Colin Meurk, Brian Molloy, Tom Moss, Colin Ogle, David Pledge, Andrew Purdie, Willy Shaw, Barry Sneddon, Bill Sykes, Geoff Walls, Josephine Ward, Colin Webb, Peter Williams, Hugh Wilson, and Anthony Wright. Each checked a proportion of the original species entries and supplied many additional comments.

Thanks are also due to Bruce Clarkson, Peter Johnson, David Massam, the late Graeme Paterson, Susan Timmins, and Peter Williams who reviewed the text and made helpful suggestions, and to Bruce Given, Catherine Harvey, John Smith-Dodsworth, Andy Thomson, the Commission for the Environment, the Department of Conservation, and the Nature Conservation Council for answering our enquiries.

Reed Methuen Publishers (now Octopus Publishing Group) kindly gave us permission to reproduce the sketches of *Dactylanthus taylorii* and *Scutellaria novae-zelandiae* and Thomas Nelson of Australia generously allowed us to reproduce the drawings of *Calochilus campestris, Chiloglottis formicifera*, and *Thelymitra matthewsii*.

We are deeply grateful to the many people who provided colour transparencies, to Pat Brooke who produced several superb drawings and to Jim Miles who prepared silhouettes of several fern fronds.

Our special thanks go to Gren Lucas for writing the Foreword, and to Graeme Platt who reviewed comments on the cultivation of many species and provided valuable additional notes.

The species entries are largely based on information in the threatened species database at DSIR's Botany Division, which has been supplied by both amateur and professional naturalists throughout New Zealand. We are greatly indebted to these people for the information they have faithfully provided over several years.

And finally, we thank the skilful and helpful staff of DSIR Publishing, especially Quentin Ruscoe and Denis Gourley, and the Director of Botany Division, DSIR, Warwick Harris, for encouragement and interest in the project.

Contents

But now ask the beasts, and let them teach you;
and the birds of the heavens, and let them tell you.
Or speak to the earth, and let it teach you;
and let the fish of the sea declare to you.
Who among all these does not know
that the hand of the Lord has done this,
in whose hand is the life of every living thing,
and the breath of all mankind?

Job 12:7-10

Dedicated to Graeme Paterson (1940-1988),
a vigorous and persistent campaigner for plant conservation.

Introduction

The need for a guide

Over 10% of New Zealand plants are recognised as being at risk. There is an urgent need to ensure that these species are adequately protected in their wild habitats. Increasingly, people in botanic and home gardens are showing interest in growing these species. It is essential that a wide range of people can recognise our rarer plants and have some knowledge of their general geographic range and the habitats where they grow. A significant number of discoveries of rare plants are made accidently rather than through deliberate surveys. We believe that publication of this guide will result in further discoveries of populations of species at risk, as well as encouraging the protection and cultivation of plants from known sites.

In selecting candidates for this guide, we have had in mind the need to revise the Native Plants Protection Act (1934) and the requirement for a guide to the species which might be listed in new legislation. The assemblage listed here reflects a consensus view on the more critically threatened plant species in New Zealand.

In most instances we hope there should be little argument over inclusion of particular species and allocation to a category of threat. Exceptions may be plants such as *Lepidium oleraceum* (Cook's scurvy grass) and *Sporadanthus traversii*. There are several healthy and unthreatened populations of *Lepidium oleraceum* on offshore and outlying islands. However, this species has almost totally disappeared on the mainland, and is vulnerable to habitat disturbance and predators. This makes it a species whose decline gives considerable cause for concern. We have included *Sporadanthus traversii* and several other species because, although they occur in very large numbers at some sites, vulnerable habitats and lack of adequate reserves make long-term survival throughout most of their range uncertain.

The Red Data Book of New Zealand (Williams & Given 1981) deals with a similar range of plants as the present volume. Of the 66 species in the Red Data Book, 20 are not included here and of the 97 taxa discussed here, 52 are not in the Red Data Book.

In all but one instance (*Teucridium parvifolium*), the exclusion of species dealt with in the Red Data Book is due to allocation to a lower conservation category which is not covered here. Publication of the Red Data Book generated considerable interest in the distribution and abundance of threatened plants and it has been inevitable that further localities have been discovered for some species. *Metrosideros carminea*, for example, was in 1981 recorded from only a few sites. Reserve surveys and assiduous searching, particularly in South Auckland and Taranaki, have revealed numerous populations of this species, some of which are in protected

The start of a pine forest near the North Cape area — but what are the losses in native species? (D. Given)

areas. However, there is still the need to adequately conserve *M. carminea*, and to monitor populations; it remains a scarce plant at many sites. Five outlying island species are found in the Red Data Book but are not treated here. Of these, two from the Three Kings Islands, *Brachyglottis arborescens* and *Alectryon grandis*, have been placed in lower categories after reassessment. *Homalanthus polyandrus* of the Kermadec Islands is showing a spectacular recovery after goats were removed. On the Chatham Islands, fieldwork and reassessment has shown that *Cotula featherstonii* and *Myrsine coxii* are not as critically threatened as previously thought.

The Red Data Book dealt only with endemic species and ignored lower taxonomic ranks. Of the taxa dealt with here but not in the Red Data Book, 32 did not qualify for the 1981 publication, either because they are varieties or subspecies (6 taxa) or because they are found native overseas (26 taxa). Four species (*Asplenium pauperequitum*, *Chordospartium muritai*, and *Metrosideros bartlettii*) have been described since 1981. The remaining 17 species are included here as a result of changes of conservation status: 12 are mainland plants, 4 are from the Chatham Islands, and 1 is from the Kermadec Islands.

Without doubt future Red Data Books and field guides will include more species. Some may be species which at present are not, or only barely threatened, but which grow in vulnerable habitats. Others will be species which have yet to be formally recognised and described. Present estimates suggest that about 10% of the New Zealand native flora is undescribed and it can be expected that a significant proportion will be plants which are extremely local in distribution.

This book describes New Zealand's most threatened plants, what they look like, the habitats they grow in, the parts of New Zealand where they may be found, and how they are being (or should be) conserved. It is not a guide to precise locations for rare plants and we have avoided closely circumscribing specific sites for species. Notes on distribution will enable the user to judge what plants occur in a particular region and, in combination with information on habitat, the likelihood of their being found at specific sites.

Many people have co-operated in producing this book and we are greatly indebted to those who have provided, and continue to supply, records for the threatened plants database at Botany Division, DSIR, and who have commented on drafts of species entries or provided photographs. The lengthy acknowledgments indicate the considerable assistance and

Leaning Rock in the Dunstan Mountains. The grasslands and fellfield of the plateau mountains of central Otago include several critically threatened plants. (D. Given)

encouragement we have received from both professional and amateur sources in New Zealand and overseas.

Types of threats

Natural catastrophes such as volcanic eruptions, floods, and fires can have profound local effects on the size and even existence of local populations of plants. Natural extinction is difficult to estimate. Nevertheless, indications are that natural extinction rates are slow compared with the accelerated rates in the last 200 years. Moreover, human interference can be both profound, affecting whole populations and communities, and widespread.

We can directly threaten species by harvesting them. In New Zealand the most obvious examples are trees such as kahikatea, rimu, and kauri which have been in high demand for many decades as sources of timber. Although plant communities dominated by these and other timber species have been greatly depleted, none of our important native timber species have yet been reduced to threatened status. Fortunately, there seems little prospect yet for development of a major cut-flower trade based on attractive native species; inevitably pressure would be put on some less common species, as has occurred in Australia and southern Africa.

More often, the habitat as a whole is affected. Some changes are fast, as when an area is flooded by a dam, a swamp is drained, or a forest is clear-felled. Other changes may be more subtle and slower. Examples include invasion by competing weeds, gradual changes in water-table levels, and creation of conditions which allow mature plants but not seedlings to survive. The insidious effects of slow habitat change may not be noticed until they have reached an advanced and perhaps irreversible state.

Sometimes habitat changes can affect a population of threatened plants from some distance. For example, pollution of a river system affects sites downstream. In protecting plant populations in their natural environment, particular attention needs to be focussed on adjacent and upstream land-use.

Remnant vegetation in the NW Ruahine Range. Many native plants are now confined to small pockets of vegetation in a sea of farmland or plantations. (D. Given)

Conservation strategies

Documentation and monitoring

The first step is to determine which species are threatened. It is important that such a list is not static. Plant distributions change and numbers may fluctuate. Botanical collecting from imperfectly known areas may increase the number of sites known for a species.

Unless the list is very short it is useful to subdivide it into categories which indicate degree and type of threat. Only species in the three IUCN (International Union for Conservation of Nature and Natural Resources) categories Extinct (Ex), Endangered (E), and Vulnerable (V) are included in the present book.

Information should be included about species, associated plants, communities in which each occurs, dispersal, reproduction, establishment, particular threats, and localities where species are found. Red Data Books are an attempt to summarise a wide range of information, but the introduction to the "New Zealand Red Data Book" (Williams & Given 1981) acknowledges that: "it is clear that there are instances when it is scarcity of information rather than scarcity of numbers that make it appear, at first sight, that certain species should be included . . . All we can hope for is that the information it contains is as accurate, and up to date as possible."

Ongoing monitoring, including inventory, survey, and detailed demographic or ecophysiological studies, will always be necessary. This allows conservation status to be reassessed and the effectiveness of protective measures to be determined.

Botany Division, DSIR, maintains site records for threatened species and uses these as a basis for developing national and regional lists. Computer-based inventories and bibliographies with on-line facilities for searching records are being developed.

Reservation

Once a species is known to be threatened, a first step should be to attempt to preserve it under natural conditions. Species are integral members of ecosystems, whose components are interdependent. Theory suggests that for many species (especially animals), if populations are at or below an effective size of about 50 breeding individuals, they may not survive long-term due to loss of genetic variability. Perhaps, more importantly, as populations become smaller they become more susceptible to extinction through fluctuations in size and through chance events. It is usually impractical to maintain large populations in conventional botanic gardens and often the only reasonable strategy is maintenance in a wild state.

Reserve notice at Ngawha Springs. Lake Waiparaheka Scientific Reserve was the first established specifically to protect a threatened species, Baumea complanata. *(D. Given)*

Relatively few New Zealand reserves exist primarily to protect individual species, although among them are some notable examples. These include Lake Waiparaheka Scientific Reserve in North Auckland, Enys Scientific Reserve in Castle Hill Basin (Canterbury), a sanctuary in Russell State Forest for *Todea barbara*, and a sanctuary for *Pittosporum dallii* in the Northwest Nelson Forest Park. However, many other protected areas established to conserve ecosystems also preserve populations of threatened plants.

New Zealand's Crown-owned protected natural areas form a network chiefly administered by the Department of Conservation. Increasingly, privately owned areas are being set aside by convenant administered mostly through the Queen Elizabeth II Trust and by the Department of Conservation.

Cultivation and gene banks

Botanic gardens and horticulture play an important role in the world-wide attempt to stem the accelerating loss of plant species. For example, the perennial teosinte of Mexico is one of the few wild relatives of maize. Plants propagated from the wild have been distributed to botanic gardens, ensuring that this agriculturally important species will survive at least in cultivation.

Where it is necessary to bring a species into cultivation because it is critically threatened in the wild, it is important to take various steps. There should be deliberate programmes to propagate it, distribute it to other gardens, and to store seed under conditions which permit long-term survival. Material should be available for research which may be critical to understanding the long-term management of a species.

An important link between *ex situ* and *in situ* conservation is the propagation of plants for reintroduction into the wild. A New Zealand example is *Hebe armstrongii*. Propagating material has been taken from wild plants, raised in cultivation and eventually replanted in the wild alongside the parent material. Similar techniques have been suggested for endemic plants of some offshore islands. Plant hygiene and quarantine are vital to prevent the introduction of predators and diseases not present on the islands.

In 1983 the Nature Conservation Council began co-ordinating conservation of threatened species through horticulture. It is hoped that attention can be drawn to those species not adequately cultivated. It is impracticable for *all* gardens to grow *all* threatened species, so a co-

Measurement of variables, such as the temperature of the hot-water Waimangu Stream, provides important information for management. (D. Given)

ordinated network will allow gardens to grow species most appropriate to their particular climate, soils, and facilities.

The availability of plants through horticulture may be a powerful weapon in combating illegal trade in endangered plants. The rarity value of some species has led to illegal trade, but there is potential to reduce rarity value by propagation and make specimens more readily available.

It must be recognised that there are some limits to the role of botanic gardens and horticulture in conservation. Attractive and colourful species are far more likely to receive attention from horticulturalists because of their greater appeal to the public. Also, plants from specialised habitats pose problems in cultivation by needing particular, and often expensive, techniques. Nevertheless, all reasonable effort should be made to cultivate threatened species of plants.

Legislation

The Native Plants Protection Act (1934) offers very limited protection for most native plant species. Recognition of serious deficiencies has led to calls for substantial revision of the act. For example, trees are not regarded as plants under the act and hence are not protected.

Other New Zealand legislation indirectly confers various degrees and types of protection. This includes the Conservation Act (1987), the National Parks Act (1980), the Reserves Act (1977), the Land Act (1961), the Town and Country Planning Act (1960), and the Forest Act (1949). Most of this legislation is territorial in nature, protecting threatened species if they occur on protected land. Often plants growing on land under Crown ownership are potentially better protected by law than those found on private land.

At the international level, the Convention on International Trade in Endangered Species of Wild Fauna and Flora (CITES) regulates trade in species at risk. The effectiveness of CITES depends largely on control of poaching and smuggling of rare and valuable species, and surveillance of imports and exports of plants. New Zealand is about to accede to the Convention.

Education

Conservation requires the support of biologists, administrators, and - most of all - the general public. People are the most powerful resource in conservation. Botanic gardens provide good opportunities for public education, allowing people to see and appreciate threatened species. But

The Chatham Islands tablelands are home to almost 40 plant species found only on these islands. (D. Given)

reserves, gardens, and even legislation are only as effective as people want them to be.

Valuable publicity focussing on needs and problems of plants at risk has been largely co-ordinated by the Nature Conservation Council. This has involved symposia and workshops, Conservation Week activities and the production of brochures, posters, and booklets explaining why species are threatened, what is being done, and how you can be involved. In 1981 the Council sponsored the Red Data Book, a basic source for data on rare and endangered plants and animals. World Wildlife Fund (N.Z.) actively supports conservation of species at risk, for example, by funding research and through the Tiritiri Matangi Island project which should provide an opportunity to recreate quasi-natural plant and animal communities of the northern offshore islands.

The educative value of legislation should not be overlooked. Providing it does not draw attention to precise sites for threatened species, it can usefully focus attention on certain species which are threatened and provoke the response: "what can I do?".

Why conserve?

Several reasons have been given in favour of conserving *all* plant species as far as is possible. Most imply preservation primarily in the wild, or at least quasi-wild state, on the assumption that species survive best among those animals and plants with which they have co-evolved throughout past millenia. Despite high initial costs (which may be necessary if land has to be purchased), conservation in the wild can, in the *long term*, also often be a cheaper option than preservation *ex situ*.

The elimination of a species from a natural community may eventually lead to loss of many other plant and animal species. Plants and animals form an interdependent network in which serious disturbance at one point can have effects throughout the whole ecosystem. We therefore need to exercise care in disturbing or manipulating natural communities. Another argument for conservation of threatened species is that the world around us is a repository of information which we have only begun to delve into. Like any library, once parts are missing, it is incomplete but, unlike a library, once our books (in this instance biological species) are lost they cannot be replaced. Others have pointed out that the full potential of the plant world is untapped, particularly for chemicals, energy, and food. This makes it important to preserve at least representative samples of all the plant world. A synthetic world, fully divorced from nature, is probably an impossible dream.

Plants are not only useful to people but are essential. Wild plants have an important traditional role in New Zealand as sources of medicines, foods, and materials for crafts. As such they are part of the New Zealand heritage. Wild plants provide valuable genes for crop breeding as well as sources of new pharmaceuticals, oils, and other desirable products.

The "right to exist" argument is essentially moral and contends that the human species has no innate right to hound to extinction other species with which it shares this planet. This argument can be extended to a "quality of human existence" argument, that a world rich in biological diversity is a *better* world to live in. Some would go further and argue that only a world in which a harmonious coexistence with nature is possible is the *right* world to live in. There may be a need to go further in developing an environmental ethic than most people have done up to the present. Paul Taylor (1986) and Bryan Norton (1987) argue persuasively for an ethic in which we recognise animal and plant species as having innate rights which we must respect. This is in contrast with the attitude, according to which the natural environment and its wild biotic communities are valued only as objects to be exploited for human use, rather than a heritage to cherish.

In a recent review of threatened habitats, ecologist Norman Myers (in Hall 1984) suggested that:

> "the surface of the earth is undergoing man-imposed changes that match the upheavals of a Pleistocene glaciation. Indeed some observers consider that the broad-scale disruption of the planet's ecosystems is as great as virtually any evolutionary dislocations that have occurred through the geologic past, and whereas the upheavals of prehistory took millions of years to run their course, the current alterations are occurring within decades, within the twinkling of an evolutionary eye."

These are strong words but they reflect an increasing concern for our use of the environment, for the degree of modification which often results, and for the accelerating rate of loss of both species and biological communities.

Some tropical ecologists suggest that virtually all lowland forests of Southeast Asia will be logged by the year 2000. Worldwide, wetlands are disappearing rapidly, particularly as they are often regarded as "unproductive". Semi-arid scrublands and grasslands readily undergo irreversible degradation through burning, overgrazing, and the continual search for wood-fuel; they are being turned to deserts at the rate of approximately 60 000 km^2 per year, an area one-quarter the size of New Zealand.

New Zealand has seen the disappearance of habitats and ecosystems, and depletion of species. Its original forest cover of over 70% has been reduced to about 25%. It has been estimated that less than 10% of our wetlands remain in good "conservable" condition.

It is, therefore, not surprising that about 10–12% of New Zealand plant species are regarded as threatened. This includes some species which may have always been very local in distribution, e.g., plants of very small islands, but many are species which at one time occupied a much wider geographic range. With about 81% of our native flora endemic (i.e., found native nowhere else in the world) we have a national responsibility to conserve our flora for future generations as well as our own. No other country is likely to take primary responsibility for this, nor should they be expected to do so.

For those who want to go further and obtain detailed information or become involved in the conservation of threatened plants, selected references and a list of addresses for further advice are included in this guide. Local botanical and biological groups, and national conservation organisations, can provide information on your region and advise on ways in which you can become involved in conserving threatened plants.

Categories of threat

Categories used in this book are based on those used by the Conservation Monitoring Centre of the International Union for Conservation of Nature and Natural Resources (IUCN) in their worldwide survey of threatened species.

Presumed Extinct (Ex): This category is used only for species which are no longer known to exist in the wild after repeated searches of the type locality and other known or likely places. It includes plants extinct in the wild but surviving in cultivation.

A plant may be listed as extinct in one country while surviving in another. For example, *Thelymitra matthewsii* is probably extinct in New Zealand but still occurs in Australia. Extinction can never be regarded as more than a probability, and rediscoveries are occasionally made.

Endangered (E): Species in danger of extinction and whose survival is unlikely if the causal factors continue. Included are those whose numbers have been reduced to a critical level or whose habitats have been so drastically reduced that they are considered to be in immediate danger of extinction.

Recovery operations involving the transfer of plants such as Hebe armstrongii *back to the wild are increasingly important. (D. Given)*

We include all species whose populations are so few or small, that loss by natural means, such inability to breed due to lack of genetic diversity or a catastrophe such as a landslip, becomes possible. We consider it is useful to include plants whose survival in the wild depends on habitat manipulation or continued management.

Vulnerable (V): Species believed likely to move into the Endangered category in the near future if the causal factors continue. Included are species of which most or all the populations are decreasing because of over-exploitation, extensive destruction of habitat, or other environmental disturbance; those with populations that have been seriously depleted and whose ultimate security is not yet assured; and those with populations that are still abundant but are under threat from serious adverse factors throughout their range.

It is sometimes difficult to draw a line between what is Endangered and Vulnerable on the one hand, and what is Vulnerable and Rare on the other. Vulnerable is essentially a dynamic category implying change and the need for active protection. Rare species may not need urgent protection although they will require monitoring.

Rare (R): Species with small world populations that are not at present Endangered or Vulnerable but are at risk. These are usually localised within restricted geographic areas or habitats or are thinly scattered over a more extensive range.

Indeterminate (I): This category is used for plants thought to be Extinct, Endangered, Vulnerable, or Rare, but for which there is insufficient information to allow allocation to a category.

Insufficiently Known (K): Plants placed here are suspected, but not definitely known, to belong to any of the above categories. There is insufficient information to be certain.

In New Zealand a category additional to those used by IUCN has been found useful. This is Local (L) to include plants not under threat but potentially threatened, and hence deserving some level of monitoring and possibly protection. Included are regional endemics, plants of potentially vulnerable habitats, and species occurring as frequent but small populations.

Code of conduct for conservation of threatened plants

1. You must obtain official permission before you take specimens of any plants from reserves, national parks, state forests, forest parks, and sanctuaries; the rights of landowners must be respected when you collect plants.
2. It is essential that you preserve the habitat and conditions in which wild plants grow. This means watching where you walk, not compacting soil, and taking care not to break off young shoots and branches.
3. As far as possible, take propagating material from plants already in cultivation. If you have to collect material from wild sources, take seed or cuttings rather than uprooting whole plants.
4. "Rescue attempts" require knowledge of the biology and habitat requirements of plants and should be undertaken only following approval by an appropriate conservation and administrative authority, and on scientific advice.
5. If a wild plant is scarce in a particular locality, take a photograph; if a voucher specimen is necessary, take the minimum needed for identification.
6. Trampling and "gardening" before taking photographs of plants may reveal the site to others. After photography, repair any damage caused by feet, knees, tripod legs, etc.
7. The exact location of critically rare wild plants should not be disclosed indiscriminately; such sites should be visited by small parties only.
8. Inform the appropriate scientific and conservation authorities if rare plants are discovered.
9. Respect requests from conservation bodies and landowners not to visit particular sites at certain times.
10. Do not introduce plants into wild habitats without first obtaining the approval of appropriate authorities and appropriate scientific advice. When native plants are introduced into wild habitats, specimens propagated from locally derived materials should be used to avoid genetic contamination.

How to use this guide

The species featured in this guide are arranged alphabetically within several broad groups, most of which were derived by combining species which are similar in appearance and growth form, or habit (e.g., climbers and scramblers). The remaining groups consist either of plants from similar habitats (e.g., aquatic herbs) or species with taxonomic affinities (e.g., orchids). An index on p. 13 details this arrangement of species. A second index (p. 15) lists species which might be encountered within a particular habitat. Some species are listed in more than one category in the "Arrangement of Species in this Book" and habitat indexes, but only one entry is given in the text. Table 1 on p. 12 gives the numbers of threatened species recorded from various habitats in New Zealand.

The entry for each species includes a photograph or line drawing for rapid preliminary identification, and a more precise description to ensure exact identification. Notes on distribution are supplemented by maps showing present known distribution. Present distribution throughout the main islands of New Zealand and beyond New Zealand is also briefly summarised in italics. Descriptions of characteristic habitats, together with the notes on distribution, enable you to determine which plants might be found in a particular habitat in a particular region of New Zealand. An outline of the conservation status of each species includes estimates of population levels and information on threats and measures taken to protect the species in the wild. Where possible, an estimate of each species' abundance in cultivation has been included, along with a statement indicating whether it has proved easy or difficult to grow. Included under "Comments" is miscellaneous information such as features distinguishing closely related species, taxonomic relationships, and an explanation of each species' scientific name.

To be as concise and accurate as possible, we have used some technical terms, especially in the species descriptions. Most of these terms are explained in the Glossary (p. 143). Diagrams explaining terms used to describe the shape of laminae (leaf blades) are included at the end of the Glossary. Sometimes the shape of a lamina, or leaf blade, cannot be adequately described by any one of these terms, as it is intermediate between two. In such instances the two terms which most closely describe its shape are both noted in the text, joined by a hyphen, e.g., elliptic-obovate.

Botanical names of species referred to in the text by a common name only can be found on p. 141.

Flowering and fruiting periods of wild populations have been included wherever these are known. However, remember that flowering and fruiting periods of both wild and cultivated plants often vary from one year to the next, and according to local climate.

Table 1: Numbers of threatened species recorded from various habitats in New Zealand.

	Presumed extinct	Endangered	Vulnerable	Total
Sandy shorelines	–	2	7	9
Rocky shorelines and stony beaches	1	1	3	5
Coastal cliffs and steep coastal slopes (including slopes covered with low scrub)	1	9	13	23
Coastal and lowland forest and tall scrub	1	10	19	30
Inland forest	–	2	7	9
Inland scrub	4	6	10	20
Grassland, herbfields, and dry marshland	1	4	7	12
Inland cliffs and rock outcrops	–	4	1	5
Limestone talus	1	1	3	5
Swamps, bogs, and seepages	–	5	17	22
Lake margins and aquatic habitats	–	–	7	7
Geothermal areas	–	2	2	4
TOTAL	9	46	96	151

Arrangement of species in this book

Habitat index

Abbreviations

2 - also listed in category 2 (or other category as indicated)
C - endemic to the Chatham Islands
K - endemic to the Kermadec Islands
PK - endemic to the Poor Knights Islands
Sn - endemic to the Snares Islands
3K - endemic to the Three Kings Islands
Ex - presumed extinct

1. Sandy shorelines

Acaena pallida
Eleocharis neozelandica
Embergeria grandifolia(C)3,4
*Euphorbia glauca*2,9
Gunnera hamiltonii
*Hibiscus diversifolius*10
Myosotidium hortensia(C)2
*Ophioglossum petiolatum*10
Theleophyton billardierei

2. Rocky shorelines and stony beaches

*Euphorbia glauca*1,9
Lepidium banksii
Lepidium obtusatum(Ex)
Lepidium oleraceum 3
Myosotidium hortensia(C)1

3. Coastal cliffs and steep coastal slopes (including slopes covered with low scrub)

Aciphylla dieffenbachii (C)
Anisotome acutifolia (Sn)4,7
Asplenium pauperequitum (PK)4
Carex ventosa (C)4
Carmichaelia fieldii
Carmichaelia williamsii 4
Celmisia adamsii
Chordospartium muritai
Embergeria grandifolia (C)1,4
Hebe elliptica var. *crassifolia*
Hebe speciosa
Lepidium oleraceum 2
Leptinella nana 7,10
Leptinella rotundata
Myoporum debile (Ex)
Olearia pachyphylla 4
Pittosporum michiei
Plantago spathulata subsp. *picta*
Pomaderris oraria var. *novae-zelandiae*
Scutellaria novae-zelandiae 5,6,9
Stilbocarpa lyallii 4
Stilbocarpa robusta (Sn)4
Wahlenbergia stricta

4. Coastal and lowland forest and tall scrub

Anisotome acutifolia (Sn)3,7
Asplenium pauperequitum (PK)3
Astelia chathamica (C)11
Boehmeria australis var. *dealbata* (K)
Calochilus campestris
Calystegia marginata
Carex ventosa (C)3
Carmichaelia williamsii 3
Christella dentata 10
Clianthus puniceus 5,6
Cyathea kermadecensis (K)
Elingamita johnsonii (3K)
Embergeria grandifolia (C)1,3
Hebe barkeri (C)
Metrosideros bartlettii 10
Myrsine oliveri
Olearia pachyphylla 3
Pennantia baylisiana (3K)
Phylloglossum drummondii 6
Pittosporum obcordatum 5,10
Pittosporum pimeleoides 5
Pomaderris apetala
Pomaderris hamiltonii 6
Pratia physaloides
Pseudopanax gilliesii
Stilbocarpa lyallii 3
Stilbocarpa robusta (SN)3
Tecomanthe speciosa (3K)
Todea barbara 6
Trilepidea adamsii (Ex)6

5. Inland forest

Chiloglottis gunnii
Chordospartium stevensonii 6
Clianthus puniceus 4,6
Dactylanthus taylorii
Hebe acutiflora
Pittosporum dallii 6
Pittosporum obcordatum 4,10
Pittosporum pimeleoides 4
Scutellaria novae-zelandiae 3,6,9

6. Inland scrub

Baumea complanata
Caleana minor
Calochilus robertsonii 12
Carmichaelia kirkii 10,11
Chiloglottis formicifera (Ex)
Chordospartium stevensonii 5
Clianthus puniceus 4,5
Corybas unguiculatus 10
Doodia aspera (Ex)
Hebe armstrongii 7
Hebe cupressoides 7
Helichrysum dimorphum
Phylloglossum drummondii 4
Pittosporum dallii 5
Pomaderris hamiltonii 4
Pomaderris phylicifolia var. *polifolia*
Pterostylis nutans (Ex)
Scutellaria novae-zelandiae 3,5,9
Todea barbara 4
Trilepidea adamsii (Ex)4

7. Grassland, herbfields, and dry marshland

Aciphylla traversii (C)10,11
Anisotome acutifolia (Sn)3,4
Carex inopinata
Hebe armstrongii 6
Hebe cupressoides 6
Lepidium kirkii
Leptinella nana 3,10
Mazus pumilio 10
Olearia semidentata (C)10
Simplicia laxa 8

Sporadanthus traversii 10
Stellaria elatinoides (Ex)

8. Inland cliffs and rock outcrops
Celmisia adamsii
Celmisia macmahonii var. *macmahonii*
Chionochloa spiralis 9
Myosotis albosericea
Simplicia laxa 7

9. Limestone talus
Chionochloa spiralis 8
Euphorbia glauca 1,2
Myosotis colensoi
Myosotis traversii var. *cinerascens* (Ex)
Scutellaria novae-zelandiae 3,5,6

10. Swamps, bogs, and seepages
Aciphylla traversii (C)7,11
Baumea complanata 12
Carex chathamica (C)11
Carmichaelia kirkii 6,11
Christella dentata 4
Cortaderia turbaria 11
Corybas unguiculatus 6
Cryptostylis subulata
Cyclosorus interruptus 12
Hibiscus diversifolius 1
Leptinella nana 3,7
Lycopodium serpentinum
Mazus pumilio 7
Metrosideros bartlettii 4
Myriophyllum robustum 11
Olearia semidentata (C)7
Ophioglossum petiolatum 1
Pittosporum obcordatum 4,5
Ranunculus urvilleanus
Sporadanthus traversii 7
Thelymitra matthewsii
Thelypteris confluens
Triglochin palustre 11

11. Lake margins and aquatic habitats
Aciphylla traversii (C)7,10
Astelia chathamica (C)4
Carex chathamica (C)10
Carmichaelia kirkii 6,10
Cortaderia turbaria 10
Hydatella inconspicua
Myriophyllum robustum 10
Triglochin palustre 10

12. Geothermal areas
Baumea complanata 10
Caleana minor
Calochilus robertsonii 6
Cyclosorus interruptus 10

trees and tall shrubs (above 2m high)

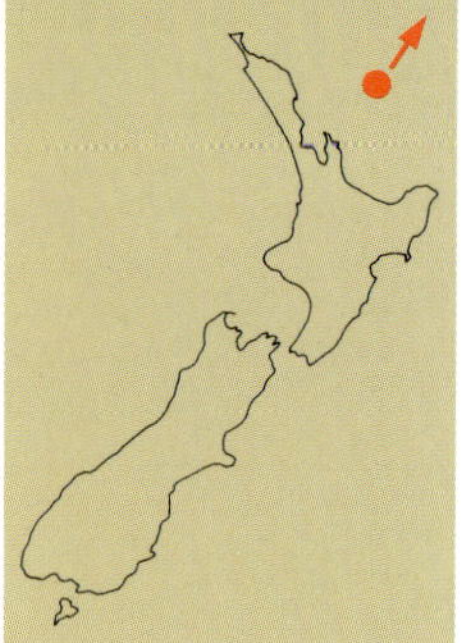

Boehmeria australis var. dealbata

Kermadec nettle-tree

Urticaceae

Description: Shrub or small tree up to 8 m. Leaves have hairy petioles up to 7 cm long, laminae are 4-19 × 2-8.5 cm wide, ovate to ovate- lanceolate, margins toothed. The lower surfaces of leaves and branchlets are densely covered in white hairs. Dense clusters of tiny, sessile, green flowers are produced in the leaf axils. Male flowers have a deeply divided perianth, are covered in stiff hairs, and are 2-3 mm long. Female flowers have a tubular perianth which is about 2 mm long and a protruding, filamentous white to pale pink stigma. They are followed by small, white ovoid achenes.
Flowers and fruit can be found throughout most of the year.

Distribution: *Kermadec Islands. Endemic. Boehmeria australis* var. *dealbata* once occurred on both Raoul and Macauley Islands, but is now found only on Raoul Island in the Kermadecs.

Habitat: This small understory tree grows in open areas and is usually found on the lower parts of Raoul, often in gullies amongst ferns.

Conservation: *Boehmeria australis* var. *dealbata* recently became very rare on Raoul Island, the few scattered populations each containing less than half a dozen trees. However, the reduction in numbers and eventual elimination of goats from Raoul in 1984 has led to the appearance of several young plants at various sites. Raoul Island is now a nature reserve. A few hundred specimens of *B. australis* var. *dealbata* are estimated to be in cultivation. It is easily grown from cuttings and seed.

J. Miles

Comments: For many years *Boehmeria australis* var. *dealbata* was regarded as a distinct species and was known as *B. dealbata*. Then, in 1977 the Kermadec plants were given varietal status on the basis of very strong similarities with *B. australis* var. *australis* which grows on Norfolk Island. Although it is a member of the nettle family, *Boehmeria australis* var. *dealbata* lacks the stinging hairs characteristic of many members of this family.

Coprosma acutifolia, Kermadec poplar *(Homalanthus polyandus),* and Kermadec five finger (*Pseudopanax arboreus* var. *kermadecensis*) have shown such prolific regeneration after goats were eradicated from Raoul Island that they are no longer considered vulnerable or endangered. These species, together with *Boehmeria australis* var. *dealbata*, provide encouraging examples of the success which can be achieved through careful management of an island reserve. *Boehmeria*: after the German botanist George Boehmer (1723-1803); *australis*: southern; *dealbata*: whitewashed, referring to the white undersurface of the leaves.

Status: Vulnerable.

Carmichaelia williamsii

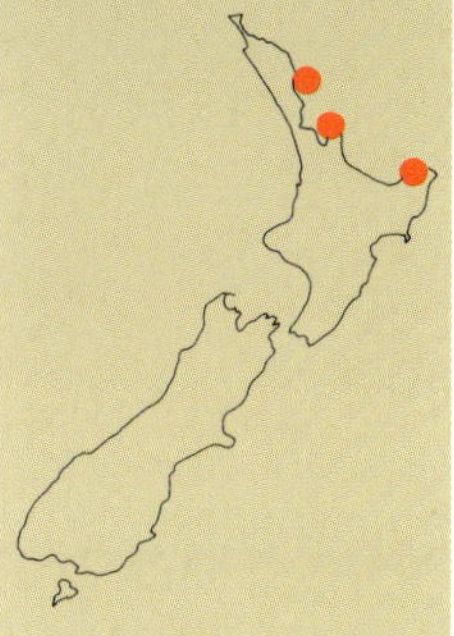

Fabaceae

Description: Erect, much-branched shrub or small tree up to 4 m with distinctive flattened and closely grooved branchlets 8-14 mm wide. Leaves only on young plants or plants growing in shady sites and consist of 1-3 narrow, inversely heart-shaped to obovate or sometimes elliptic leaflets up to 2.5 cm or more. Flowers, 1-6 in a cluster, up to 30 mm long. Their petals are yellowish, with purple veins and blotches. The dark brown to black seed pods are up to 30 mm long, each with 6-15 red seeds which may have black markings. Flowers and seed pods are produced throughout the year.

Distribution: *North Island, Poor Knights Islands. Endemic.* Small populations of *Carmichaelia williamsii* survive on the Poor Knights Islands, Little Barrier Island, and several other offshore islands in the Hauraki Gulf and near Coromandel Peninsula. On the mainland the species is known from only a few North Island sites in the Eastern Bay of Plenty and near East Cape.

Habitat: *Carmichaelia williamsii* is characteristically found in open forest on steep coastal sites.

Conservation: Less than 500 plants of this species are estimated to remain in the wild and many colonies consist of only a few plants. It seems that the present range of the species is a relict distribution and that *Carmichaelia williamsii* is still declining rapidly. Mainland colonies have probably been disturbed and prevented from regenerating by habitat destruction and browsing by introduced animals such as sheep, cattle, and goats. The Poor Knights Islands and Little Barrier Island are Nature Reserves. *C. williamsii* is easily grown from seeds and cuttings and is widely cultivated in New Zealand. It is sometimes available commercially.

Comments: This attractive species is closely related to *Carmichaelia exsul* of Lord Howe Island. It is the only yellow-flowered species of *Carmichaelia* in New Zealand. *Carmichaelia:* after Dugald Carmichael (1772-1827), a Scottish army officer who was stationed in South Africa and who wrote an account of Tristan da Cunha; *williamsii:* after William Williams (1829-1916), the third Bishop of Waiapu, who discovered this species at Hicks Bay.

Status: Vulnerable.

D. Given

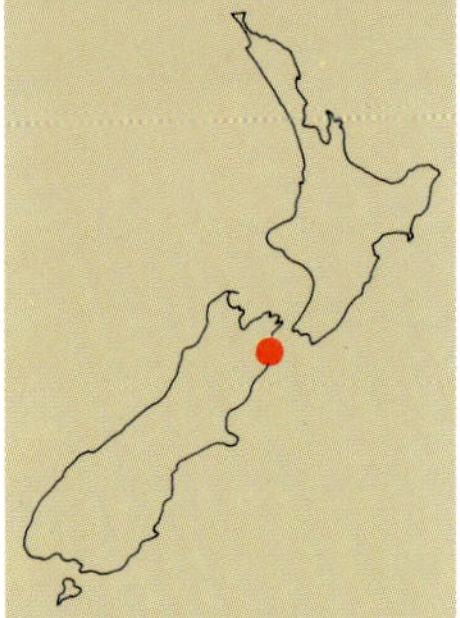

Chordospartium muritai

Clifford Bay broom

Fabaceae

Description: Leafless tree up to 6 m. Trunk up to 25 cm in diameter, with a compact canopy of branchlets, the uppermost erect, lower ones pendulous. Mature branches erect and greyish brown with circular scars near the points of branching. Actively growing branches yellowish green with brown or grey markings and a distinctive striped pattern. All branchlets are striped, deeply grooved, and flattened to round in cross section. Flowers in erect cylindrical clusters up to 7 cm long, each about 4 mm long, petals whitish with purplish violet markings near the base and violet veins. Individual trees do not flower every year. At maturity the seed pods are hairy, obliquely obovate, about 3.5 mm long, including a stout, blunt beak about 1 mm. Pods usually contain a single, yellowish green seed. Flowers: December-January; Fruit: February-May.

Distribution: *South Island. Endemic.* The only known population grows at Clifford Bay in Marlborough.

Habitat: Plants grow on a deeply eroded coastal cliff of coarse, free-draining gravel.

Conservation: A total of 28 trees are known to survive in the wild. These plants are threatened by erosion of the cliff on which they grow and the possibility of fire. Natural regeneration at the site is uncommon and is limited by the browsing of rabbits and the spread of weeds. Periodic grazing by sheep has also prevented the past establishment of seedlings here.
Young plants are being replanted at the site, which is now included in a scientific reserve. Propagation and germination characteristics of this species are the same as those of *Chordospartium stevensonii.*

Comments: *Chordospartium muritai* was described in 1985, in the *New Zealand journal of botany. Chordospartium:* from *chorda*, the string of a musical instrument, and *spartos*, a broom; *muritai:* sea breeze, alluding to its coastal habitat.

Status: Endangered.

D. Given

Chordospartium stevensonii

weeping broom, cord broom

Fabaceae

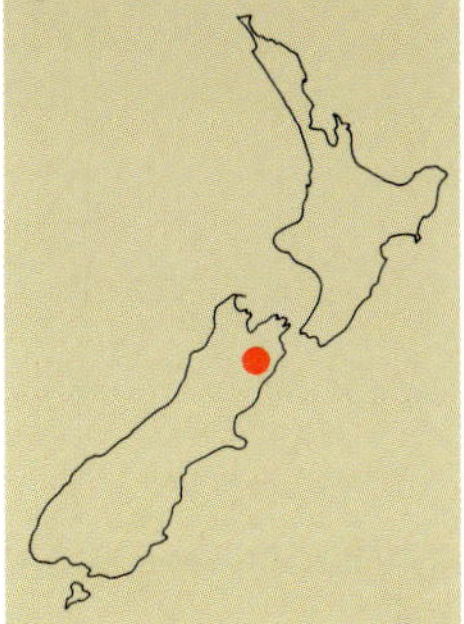

Description: Leafless tree up to 8 m with an umbrella-like canopy of pendulous branchlets, trunk up to 30 cm in diameter. Mature branches greyish green and irregularly ringed with circular scars which seem related to points of branching. Actively growing branches have very distinctive striped bark, and terminate in green, deeply grooved cylindrical branchlets about 2 mm in diameter. Flowers in pendulous cylindrical clusters up to 8 cm long, each flower 5-10 mm long, pale lavender with darker veins on the broad upper petal. Hairy, diamond-shaped to suborbicular pods are about 6 mm long and contain 1-3 yellowish green seeds.
Flowers: December-February; Fruit: February-May.

Distribution: *South Island. Endemic. Chordospartium stevensonii* grows only in Marlborough, near tributaries of the Clarence and Awatere Rivers.

Habitat: Weeping broom usually grows in free-draining colluvial or alluvial soils and seems to be intolerant of deep shade.

Conservation: In the past *Chordospartium stevensonii* may have been relatively common in the river valleys of Marlborough, but it is now found in only a few localities in inland Marlborough. Although several hundred trees grow above the timberline in the Seaward Kaikoura Range, their mountain habitat is vulnerable to erosion and fire. Less than 100 plants have been fenced off in a protected private land reserve in the Jordan River Valley. This attractive tree is frequently found in public gardens, chiefly in New Zealand, and less often in private gardens. It can be propagated from hardwood cuttings but the success rate is low. Fresh seed germinates readily but old seed germinates slowly unless it is pre-treated by cutting or soaking. In their second and third years young plants look very much like dead sticks.

Comments: *Chordospartium stevensonii* was in cultivation before it was "officially" discovered. Humphrey Weld, while a guest of Mr C.G. Teschemaker at Avon Valley in 1864, sent seed to his home in Bridgeport, England, where plants were raised. It was flowering there well before Thomas Cheeseman described it in 1911. Individual trees do not flower every year.
Chordospartium: from *chorda*, the string of a musical instrument, and *spartos*, a broom; *stevensonii*: after George Stevenson who discovered this plant near the mouth of the Clarence River early this century.

Status: Vulnerable.

D. Given

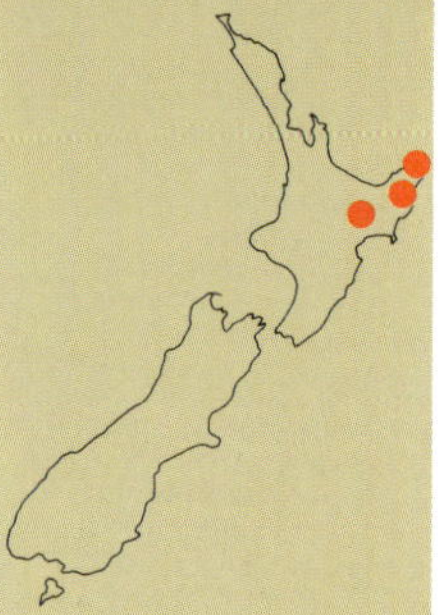

Clianthus puniceus

kowhai ngutu-kaka **kaka beak**

Fabaceae

Description: A much-branched, woody shrub usually up to about 3 m but sometimes taller. Leaves up to 15 cm, with as many as 15 pairs of narrowly oblong leaflets each 1.5-2.5 cm long. Large scarlet flowers are 5-8 cm long and in pendulous clusters of 15-20 flowers. Seed pods black, up to 8 cm, and opening widely to reveal numerous, small kidney-shaped seeds surrounded by dense white hairs.
Flowers: July-December; Fruit: December-February (or April in cultivation).

Distribution: *North Island. Endemic.* Approximately 14 colonies of *Clianthus puniceus* are known from near Lake Waikaremoana, 2 more occur along the Ruakituri and Waiau Rivers and it occurs near Tolaga Bay, Anaura Bay, and East Cape. A few plants reputed to be var. *maximus* have been discovered near East Cape, and between East Cape and Gisborne. The earliest collections of *C. puniceus* were made near Rotorua and in coastal localities in the North Island, but it may have been deliberately cultivated by Maoris in many of these places.

Habitat: Kaka beak occurs in a range of habitats, including bluffs, coastal cliffs, river and lake margins, and successional shrublands.

Conservation: The decline in numbers of *Clianthus puniceus* is probably a result of habitat destruction and predation by introduced herbivores such as goats, deer, and possibly the Australian possum. Some wild populations are within Urewera National Park but total less than 150 plants. Several plants resembling var. *maximus* grow in a proposed scenic reserve. Another three plants grow on an adjoining property.

Kaka beak is an outstanding garden shrub and is extensively cultivated both in New Zealand and overseas. Plants are readily propagated from seed and cuttings but are relatively short-lived and very susceptible to defoliation by leaf miner and to the development of unsightly "witches broom" outgrowths on the stems, caused by minute eriophyid mites.

J. Rolfe

Comments: White- and pink-flowered forms are found in cultivation. The white-flowered forms are reputed to have occurred on the Tiniroto cliffs. Colenso's var. *maximus* forms a small tree up to 5 m, with darker flowers and larger leaflets; its status is uncertain. *Clianthus:* splendid flower; *puniceus*: purple-red; *maximus*: largest.

Status: Endangered.

Elingamita johnsonii

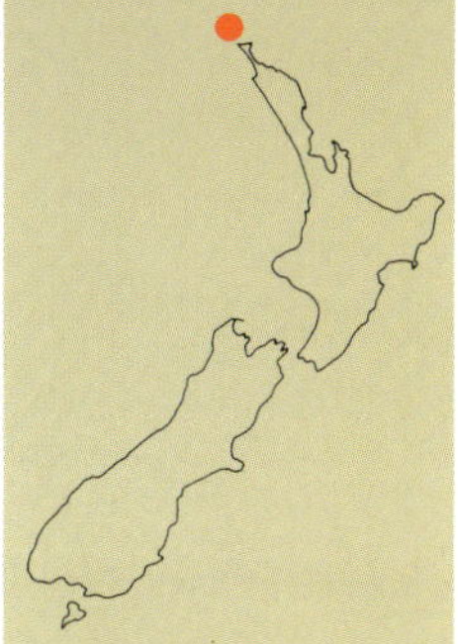

Myrsinaceae

Description: Tree up to 4 m, canopy spreading up to 5 m in diameter, bark smooth, pale grey. Glossy, gland-dotted leaves have leathery, obovate to broadly obovate laminae 9-18 x 4-9 cm, margins entire. Leaves have short, stout petioles about 1 cm long. Yellowish tubular flowers are grouped in short clusters at tips of stems, each flower about 2.5 mm in diameter and supported by a short stalk. Fruit, globose, drupe-like, up to 17 mm in diameter, and changing from green to a distinct bright red as it ripens.
Flowers: October-December (March?) in cultivation; Fruit: drupes on cultivated plants ripen 12-15 months after flowering.

Distribution: *Three Kings Islands. Endemic.* For over 20 years *Elingamita johnsonii* was thought to be confined to West Island. Then, in 1983, a colony was discovered on nearby Hinemoa Rock in the Princes Islands.

Habitat: *Elingamita johnsonii* is widespread in forested areas of West Island.

Conservation: Early observations of this plant and its habitat were made in dense fog and for many years it was thought that no more than "perhaps a dozen trees" grew on the island. In 1983, however, numerous trees and seedlings were found on West Island. No flowers were seen but bunches of ripe red fruit were abundant.

Elingamita johnsonii is cultivated in several parts of New Zealand and overseas. It can be grown from cuttings and seed. The seed may take 10 months or more to germinate. Young plants are killed by frosts of only a few degrees.

Comments: *Elingamita johnsonii* was discovered by Major Magnus Johnson when he landed on West Island in January 1950. He returned to the island in the following year, accompanied by Professor Geoffrey Baylis who later described the new species. No further landings were made until 1982. Although it is superficially similar to karaka when not in fruit, *Elingamita johnsonii* has slightly lighter green leaves and very thick, fleshy branchlets. It is the only species in the genus. *Elingamita*: commemorating the steamship *Elingamite* which was wrecked in 1902 near the foot of the cliffs where this plant was first discovered; *johnsonii*: after Major Magnus Johnson.

Status: Vulnerable.

A. Wright

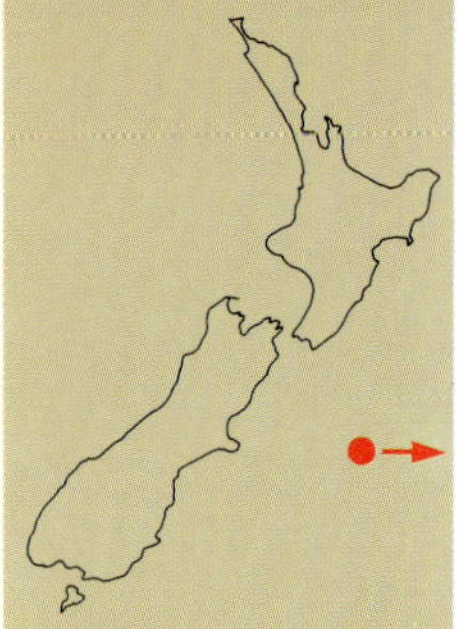

Hebe barkeri

Barker's koromiko

Scrophulariaceae

Description: Stout, upright shrub or small tree up to 7 m. Fleshy, pale green, lanceolate or oblong-lanceolate leaves are 2.5-12 x 0.7-4 cm. Fine hairs covering both surfaces of the leaves give them a velvety texture. Leaf buds have no sinus gap between leaf bases. Numerous pale mauve to white flowers are borne in elongate inflorescences up to 9 cm long, each flower about 6 mm long with tiny hairs covering its style, ovary, and calyx. Ovoid seed capsules are about 5 x 4 mm. Flowers: December-March; Fruit: January-April.

Distribution: *Chatham Islands. Endemic.* Several large populations of this species have been discovered on Chatham Island and a few smaller populations also occur on Pitt Island and South East Island.

Habitat: *Hebe barkeri* occurs in scrub and forest and once it was probably an important constituent of the lowland and tableland forest on Chatham Island.

Conservation: Destruction of habitat and grazing of young plants has resulted in a decline in the abundance and range of *Hebe barkeri*. Several hundred plants grow in a reserve on Pitt Island, and the species is also found in reserves on Chatham Island, South East Island, and Glory Island but the precise number of plants in these reserves is not known. A few hundred specimens of *H. barkeri* are in cultivation. It is easily grown from cuttings and fresh seed; seed kept for more than 1-2 years often fails to germinate.

Comments: *Hebe*: the Greek goddess of youth; *barkeri*: after S.D. Barker who collected plants on the Chatham Islands and the west coast of the South Island.

Status: Vulnerable.

O. John

Metrosideros bartlettii

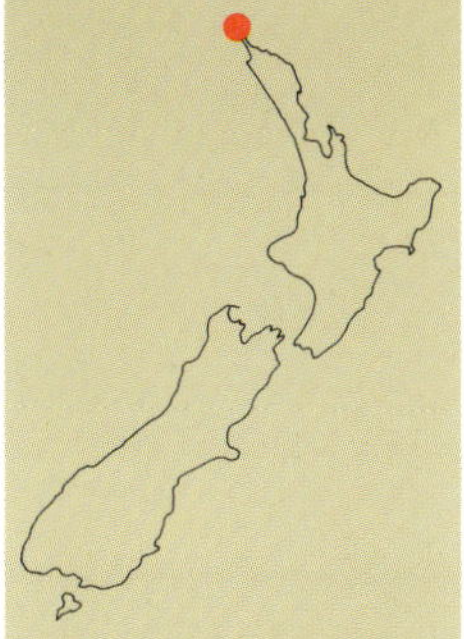

Myrtaceae

Description: Large tree up to 25 m, trunk up to 1 m or more in diameter. Bark distinctive pale grey to whitish, soft and spongy, separating into thin flakes. Leaves pale green, petiole up to 5 mm, lamina elliptic to ovate 3-4.5 x 1.5-2.3 cm. Sharply pointed tip of lamina is often elongated and twisted. The small flowers have white petals up to 3 mm and numerous stamens up to 9 mm. Old fruits retain their sepals and have strongly exserted seed capsules protruding up to 2.5 mm. Seed has not yet been described.

Distribution: *North Island. Endemic.* This species was discovered in 1975 near Cape Reinga, at the northern tip of the North Island.

Habitat: Few mature plants have been found, all of which grow alongside streams or swamps within valley-head forest remnants. Numerous seedlings have also been discovered. These are epiphytic on trees or tree ferns.

Conservation: Neither of the two known colonies are reserved. Destruction of the colonies by fire or browsing by possums is a high risk. A few plants are in private gardens. Over 100 plants are being raised in a commercial nursery but this material is from less than 5 parent plants and, therefore, has little genetic variability.

Comments: *Metrosideros bartlettii* was described in the *New Zealand journal of botany* in 1985. No other tree species of *Metrosideros* has similar bark, and the only other tree-forming *Metrosideros* with small white flowers is *M. oreomyrtus*, which grows in New Caledonia. *Metrosideros*: iron wood, referring to the hardness of the heartwood; *bartlettii*: after John Bartlett (1945-1986), who discovered this species; he was a secondary school teacher and a knowledgeable plant collector.

Status: Endangered.

N. Clunie

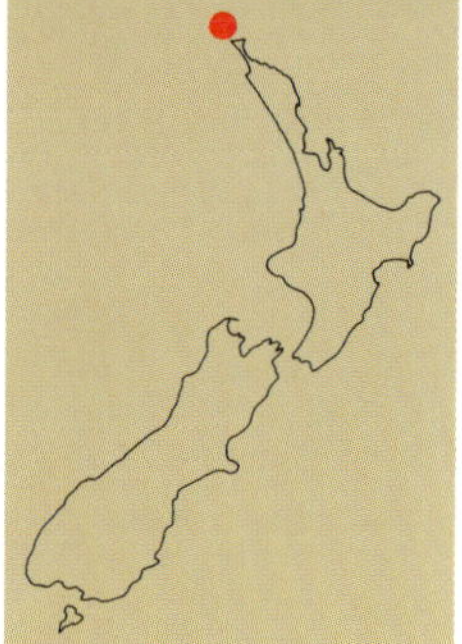

Myrsine oliveri

Myrsinaceae

Description: Shrub or small tree up to 6 m, with stiff, widely spreading branchlets and smooth, greyish bark. Leaves glossy and dark green, petioles flat and 5-10 mm long, laminae obovate to elliptic 3-10 x 1.5-4 cm. Their margins are irregularly toothed towards the tip. Juvenile leaves are scabrid and yellowish. Dense clusters of small, inconspicuous flowers are produced along the branchlets, male and female flowers being on separate plants. Fruit flattened, subglobose, drupe-like, 5-7 mm in diameter, and bright purple. Flowering and fruiting periods not known. Ripe and unripe fruit has been found in December.

Distribution: *Three Kings Islands. Endemic. Myrsine oliveri* has been found only on Great Island.

Habitat: This species grows in the low coastal forest of Great Island and is most often found in shallow coastal valleys and on steep scree slopes.

Conservation: *Myrsine oliveri* faces an uncertain future in the wild. The total population is very small, perhaps only 20 or 30 plants, and these plants are widely dispersed on Great Island. This distribution pattern endangers the species' survival as both male and female plants need to be nearby to produce fertile seed. An expedition to Great Island in 1983 discovered only two ripe fruit. Cuttings collected during the expedition failed to establish and only one seed germinated. This seedling is the only specimen of *Myrsine oliveri* known to be in cultivation.

Comments: *Myrsine*: the Greek name for myrtle or myrrh; *oliveri*: after Walter Oliver (1883-1957), one of New Zealand's greatest naturalists. He was Director of the Dominion (now National) Museum from 1928 to 1947.

Status: Endangered.

A. Wright

Olearia pachyphylla

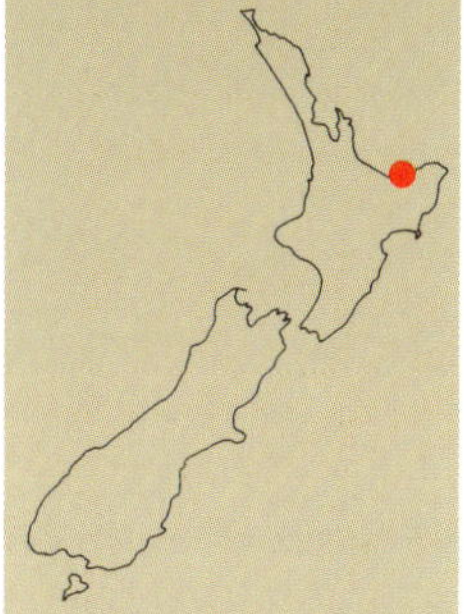

thick-leaved tree daisy

Asteraceae

Description: Spreading shrub up to 3 m, with flaking greyish brown bark and stout, grooved branchlets densely covered in brownish hairs when young. Leaves have thick, leathery laminae 5-14 x 3-8.5 cm, obovate to ovate-oblong with slightly undulate margins. Their lower surfaces are covered in a thin layer of very short white hairs. Numerous small white flower-heads are grouped in large flattened clusters towards branchlet tips. Each flower head has a characteristic cylindrical basal portion covered by 35-40 small hairy bracts. Fruits are achenes 2-4 mm long, crowned by silky, white hair-like bristles. Flowers: January-April; Fruit: March-May.

Distribution: *North Island. Endemic.* The thick-leaved tree daisy is confined to a few sites in the Bay of Plenty.

Habitat: Most plants known in the wild at present grow amongst low scrub on coastal cliffs. Some have also been found along stream banks.

Conservation: Several hundred to one thousand plants of this species are estimated to remain in the wild. A population near Opape is increasing in numbers but encroaching gorse and woody shrubs could choke out young plants, and there is also a very real threat of erosion in the area. At least one population needs to be managed to preserve the species. The expansion of a scenic reserve near Opape may ensure this. *Olearia pachyphylla* is not widely cultivated. It is readily grown from seeds and cuttings but cultivated plants are often killed by collar rot.

Comments: *Olearia pachyphylla* is one of several closely related species found along the coast of the North Island. It has been suggested that a few plants discovered near Coromandel in the late 1970s may have originated from seeds or plants taken from the Opape region by Maoris in the early 1700s when they migrated by canoe to Howick and Tamaki, stopping near Tauranga and Coromandel on the way. *Olearia:* after Johann Gottfried Oelschlaeger (1635-1711); *pachyphylla:* with thick leaves.

Status: Vulnerable.

R. Lamberts

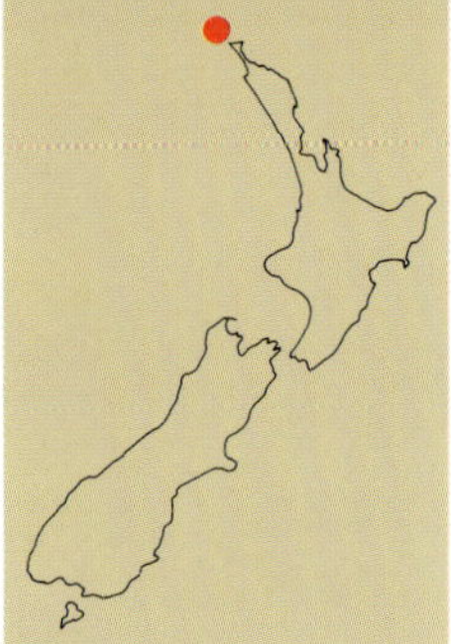

Pennantia baylisiana

Icacinaceae

Description: Small tree up to 5 m with pale brown to white bark. Branchlets have swollen leaf nodes and conspicuous corky spots on the bark which function as pores. Leaves glossy, usually clustered at branchlet tips, petioles stout and 2-4 cm, laminae are up to 28 x 16 cm and obovate with a rounded or shallowly notched tip, a tapering or slightly rounded wedge-shaped base, and entire, recurved margins. Small pits occur at the junction of mid and lateral veins on the lower leaf surface. Flowers tiny in panicles 10-15 cm long and arising below the leaves or at tips of stems. Flowers have both stamens and an ovary, but the pollen is usually not well developed. The only "drupes" which have been seen are ovoid, greenish, and up to 6 mm long. It is not definitely known whether these "drupes" were fertilised or not. Flowering and fruiting periods not known.

Distribution: *Three Kings Islands. Endemic.* This species, which is represented by only one known tree, is confined to Great Island in the Three Kings group.

Habitat: The solitary tree grows amidst coastal forest on a steep scree slope which is covered by a dense growth of ferns and herbs.

Conservation: *Pennantia baylisiana* is one of very few species which have failed to show a strong recovery in numbers since goats were removed from Great Island in 1946. One of two cuttings planted alongside the parent tree in 1971 has survived and had doubled in size by 1982. Although the long-term survival of *Pennantia baylisiana* in the wild is not certain, the species is becoming increasingly popular as a garden plant and is grown in botanic and private gardens in New Zealand. Plants can be propagated from cuttings taken from second-year wood and are sometimes available commercially. They hybridise readily with *P. corymbosa* to produce plants resembling a large-leaved form of *P. corymbosa*.

A. Wright

Comments: The single specimen remaining on Great Island is listed in the *Guinness Book of Records* as one of the world's rarest plants. The species was originally placed in the Anacardiaceae as the sole member of the genus *Plectomirtha*. In 1970 it was transferred to the genus *Pennantia* and included under *P. endlicheri*, a Norfolk Island endemic, but in 1977 it was reinstated as a distinct species on the basis of obvious morphological differences between the Norfolk Island plants and the Three Kings tree. It has been suggested that small amounts of pollen present in some flowers may sometimes fertilise another flower, thus producing the larger "fruit" occasionally found amongst bunches of small unfertilised ovaries. Most of the ovaries in each bunch are unfertilised and the whole cluster is probably aborted before the larger "fruit" matures. *Pennantia*: after Thomas Pennant (1726-1798), a Scottish naturalist; *baylisiana*: after Geoffrey Baylis (born 1913), the first Professor of Botany at the University of Otago, who discovered this species in 1945.
Status: Endangered.

Pittosporum dallii

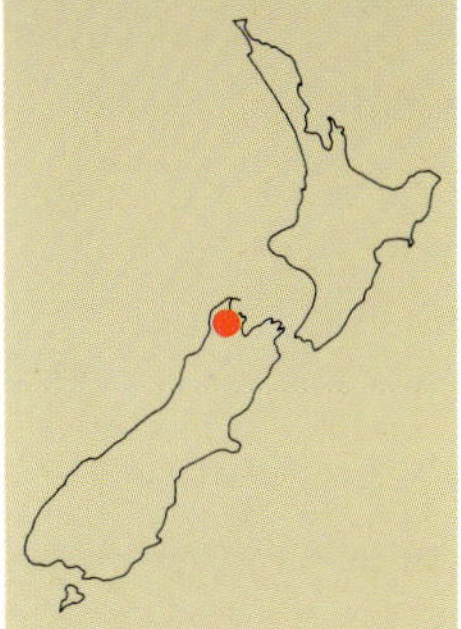

Dall's pittosporum

Pittosporaceae

Description: Large shrub or small, rounded tree up to 6 m, trunk up to 20 cm in diameter, mature bark grey and that on branchlets deep reddish to purplish black. Leaves are concentrated near the tips of branchlets. Petioles stout up to 2 cm, laminae elliptic to elliptic-oblong and 5-12 x 2.5-5.5 cm, margins usually sharply and coarsely toothed but may sometimes be more or less entire. Flowers fragrant yellowish white or less often white, 1.5-2 cm in diameter, borne in dense clusters at stem tips. Fruit a woody, ellipsoid-oblong capsule 1-1.5 cm long. Each capsule has 2 valves and contains a cone-shaped mass of dark red seeds immersed in a brownish pulp and enclosed within a papery endocarp. Endocarp and seeds may remain for a long time after the valves have been shed.
Flowers: November-January; Fruit: usually ripens over about 12 months, becoming mature in January-March.

Distribution: *South Island. Endemic.* This species is known from several areas in the Tasman Mountains, Nelson region. It has probably always been rather locally distributed.

Habitat: Dall's pittosporum has been found alongside rocky forested creeks, among boulders on forest margins, on rocky outcrops and forested cliffs, in both dense and open beech forest and in sub-alpine scrub. The species appears to be restricted to particular rock types.

Conservation: At least one thousand plants are known in the wild, where they are browsed by goats and possibly deer. At most sites there are few seedlings. Regeneration seems to be generally restricted to sites inaccessible to feral animals. Most plants are in the Northwest Nelson Forest Park, some in a forest sanctuary. *Pittosporum dallii* is in limited cultivation in New Zealand although it is highly regarded as a horticultural plant in Britain. It can be grown from cuttings and seed; however, cuttings may be difficult to root and seed can take up to 12 months or more to germinate. Flowers are rarely produced in cultivation and good seed-set seems rather sporadic, even in the wild.

B. Given

Comments: *Pittosporum dallii* has strongly sweet-scented flowers and is the only New Zealand *Pittosporum* with regularly toothed leaves (although teeth may be infrequent on some cultivated plants). *Pittosporum*: tarry or sticky weed, referring to the resinous pulp surrounding the seeds; *dallii*: after James Dall (1840-1912), a botanical collector who established a nursery at Rockville near Collingwood in the 1890s.

Status: Vulnerable.

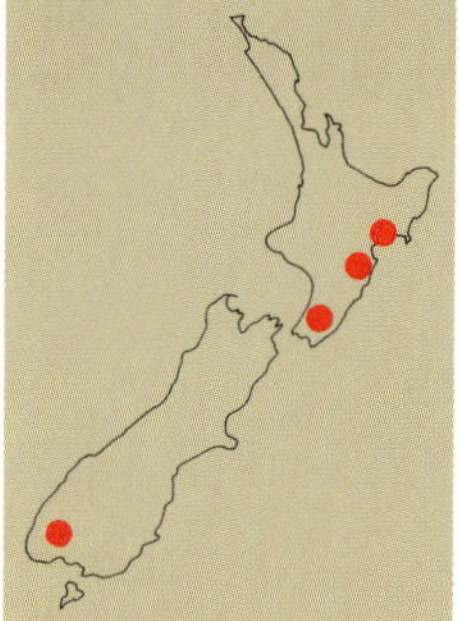

Pittosporum obcordatum

heart-leaved kohuhu

Pittosporaceae

Description: Tall shrub or small erect tree up to 5 m, with slender, divaricating and interlacing branchlets. Leaves of juvenile plants are different from those of adults, up to 2.5 cm or more long, have broad petioles, and vary from linear to spathulate with entire to lobed margins. They are dark green on the upper surface and have an obscure pattern of pale brownish blotches on the petioles and lower leaf surface. Adult plants have leaves up to 2 cm long, suborbicular to broadly obovate with entire to shallowly lobed margins; they are lighter in colour than the juvenile form. Flowers about 15 mm in diameter, in small clusters of 1-5, petals yellow and often edged with red. Seed capsules ovoid, 2-valved and about 5 mm long.

Juvenile plants of variety *kaitaiaensis* differ from var. *obcordatum* in having slightly larger leaves with sparse white hairs on both surfaces and a fringe of similar hairs along the margins. Adult plants of this variety have broadly obovate leaves which are 6-15 x 6-11 mm, seed capsules are 6-7 mm long.

Flowers: October-December; Fruit: November-January.

Distribution: *North Island, South Island. Endemic.* *Pittosporum obcordatum* may be a spectacular example of the depletion of a New Zealand species. It was first recorded from Akaroa, in the South Island, in 1840 but was never seen there again. The first North Island population was found near Kaitaia early this century. These plants were slightly different from those found previously and were therefore given varietal rank (var. *kaitaiaensis*). Plants similar to those discovered at Akaroa were found in 1920 near Wairoa, in northern Hawke's Bay, but by then the Kaitaia population had disappeared. Other populations discovered since at Kiokio, Opotiki, and Tahunga near Gisborne have also vanished. *Pittosporum obcordatum* is now known only from near Wairoa and the Tukituki River in Hawke's Bay, the Gladstone district near Wellington, and a recently discovered South Island population near Lake Manapouri in Fiordland.

E. Cameron

Habitat: All records of *Pittosporum obcordatum* are from shrubland and lowland podocarp forest on alluvial or swampy soils. The species seems to prefer river terraces and flats, especially areas which are swampy or subject to flooding.

Conservation: The sporadic distribution of *Pittosporum obcordatum* suggests that this species may have been far more widespread before the destruction of much of New Zealand's lowland podocarp forest and shrubland on alluvial soils. Probably only several tens of individuals remain in the wild in the North Island. The Manapouri population, which is within Fiordland National Park, has been estimated to consist of about 1000 adult plants, but few seedlings have been found here. The plants at Wairoa are within a nature reserve, but their long-term survival seems doubtful as the reserve is extremely small and competition with encroaching weeds may eliminate the colony. Both *Pittosporum obcordatum* and *P. obcordatum* var. *kaitaiaensis* are occasionally found in cultivation. They are difficult to establish from cuttings but can be raised from seed when good seed is ➧

Pomaderris apetala

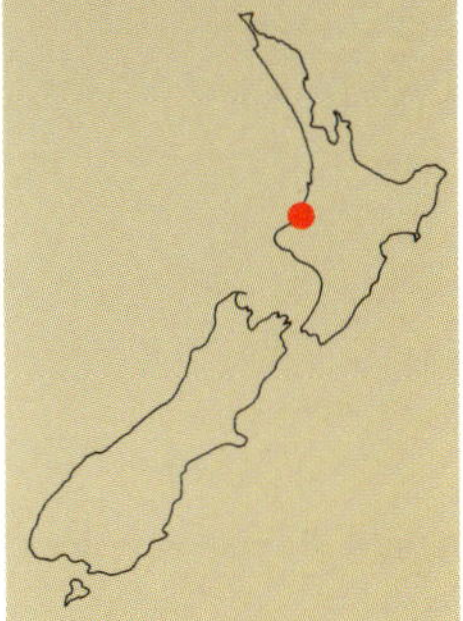

Tainui, nonokia

Rhamnaceae

Description: Erect, much-branched shrub or small tree up to 5 m tall. Leaves have narrow to broadly elliptic laminae, 5-7 x 2-3 cm, margins wavy or fine, sometimes slightly recurved, upper surface dark green, wrinkled, and with scattered, star-shaped clusters of hairs, lower surface obscured by star-shaped clusters of dense white hairs and scattered, rust-coloured hairs. Flowers pale green, lacking petals, about 6 mm in diameter, borne in open brownish, pyramidal clusters up to 20 cm long. Ripe seed capsules are half-immersed in the calyx-tube which retains its sepals and is covered in hairs similar to those of the lower leaf surface. Fruit a blackish globose capsule, about 2 mm in diameter, and containing 3 cocci each enveloping a single dark seed about 1.5 x 1 mm.
Flowers: usually November-February; Fruit: usually January-March, although flowering and fruiting periods can vary greatly from year to year.

Distribution: *North Island. Widespread in Australia.* Tainui is known from only two wild colonies in northern Taranaki. Garden escapes have been found near Nelson, Kaikoura, on Banks Peninsula, near Invercargill (all South Island), and also on Stewart Island.

Habitat: The largest remaining wild population of this coastal species grows on the sunny slopes of a shallow coastal gully in light sandy soils.

Conservation: Several hundred wild plants are within a scenic reserve and a second reserve containing less than 50 trees is being negotiated. Tainui grows readily from cuttings and seed. It is widespread in cultivation, although many garden plants may have originated from Australian material.

Comments: *Pomaderris apetala* is often confused with hazel pomaderris (*P. aspera*), an Australian species which is cultivated around Auckland and has become naturalised in a few areas. Hazel pomaderris differs from *P. apetala* in that the hairs on the lower surfaces of its leaves are not as dense and, therefore, the green lower surface of each leaf is clearly visible. Its leaves are not hairy on the upper surface. According to a Maori legend, *P. apetala* was brought to New Zealand from Polynesia on the canoe "*Tainui*" which is reputed to have landed at Mokau, the plants in the area supposedly originating from the flooring timbers of the canoe. This account of the species' origin seems unlikely, as it has never been found in the Pacific Islands. *Pomaderris*: from *poma*, lid, and *derris*, skin, referring to the membranous lid on the inner wall of each coccus which opens to release the seed; *apetala*: without petals.

Status: Vulnerable.

J. Braggins

---

available. Although germination may not occur for up to 12 months, plants usually survive once they become established in cultivation.
Comments: *Pittosporum:* tarry or sticky seed, referring to the resinous pulp surrounding the seeds; *obcordatum*: inversely heart-shaped, the shape ascribed to the leaves in the first description of plants from Akaroa.
Status: *Pittosporum obcordatum* - Vulnerable.
P. obcordatum var. *kaitaiaensis* - Presumed extinct.

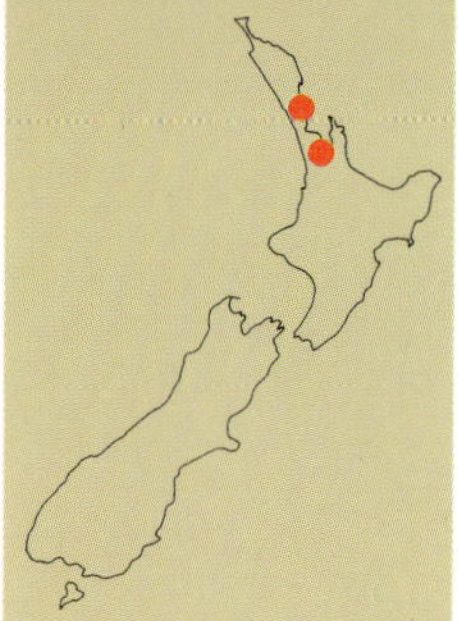

Pomaderris hamiltonii

pale-flowered kumarahou

Rhamnaceae

Description: A slender, erect, much-branched shrub or small tree, up to 3-4 m. Leaves with elliptic laminae, margins entire and recurved, usually 2-8 x 1-3.5 cm, sometimes up to 14 x 6.5 cm. Minute hairs are scattered along the midribs of the upper leaf surface and dense white to grey hairs are arranged in star-shaped clusters on the lower surface. Flowers pale, creamy yellow, 3-4 mm in diameter, in rather open, conical clusters. Each ripe seed capsule is half-immersed in a calyx-tube clad in star-shaped clusters of white hairs, and does not retain its sepals. Capsules about 3 x 2.5 mm, and containing 3 cocci each enveloping a single, shiny black seed about 1.5 x 1 mm.
Flowers: September-October; Fruit: November-January.

Distribution: *North Island. Endemic.* Several small populations of pale-flowered kumarahou occur from Warkworth to Omaha Bay and two further colonies grow on the west coast of the Firth of Thames.

Habitat: Pale-flowered kumerahou is usually found amongst scrub on roadside banks.

Conservation: No deliberate measures have been taken to protect any of the known populations which altogether may total several hundred to about one thousand plants. It is locally abundant, but the species occupies a vulnerable roadside habitat; plants have occasionally been destroyed by gangs "tidying" roadsides in Rodney County. Although only a few specimens seem to be in cultivation, pale-flowered kumarahou can be readily grown from seed.

L. Moore

Comments: Kumarahou (*Pomaderris kumeraho)* on the right of the illustration is similar to *P. hamiltonii* on the left but has brighter yellow flowers arranged in tight, round clusters. The dark brown bracts which surround the early flower buds until late autumn are wider and darker in *P. kumeraho* than in *P. hamiltonii*. Pale-flowered kumarahou was described in 1961.
Pomaderris: from *poma*, lid, and *derris*, skin, referring to the membranous lid on the inner wall of each coccus which opens to release the seed; *hamiltonii*: after the Hamilton family of Warkworth.

Status: Vulnerable.

E. Cameron

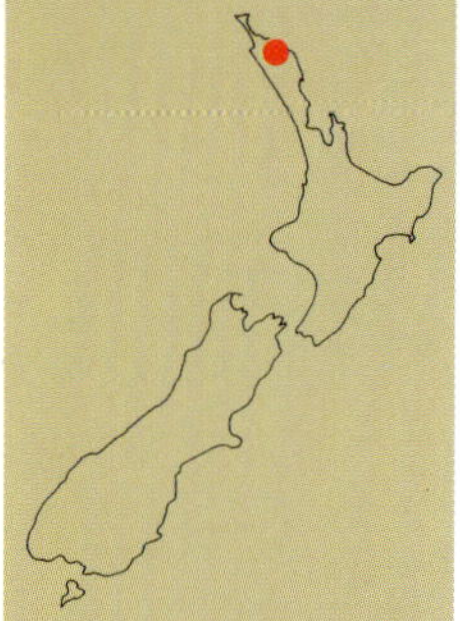

Pseudopanax gilliesii

Araliaceae

Description: Shrub or small tree growing up to 5 m tall. Leaves have a petiole up to 13.5 cm long, laminae each consist of either a single ovate leaflet 4-8 cm long with sharply, toothed margins, or 3 lanceolate to narrowly ovate leaflets each 5-11 cm long. Flowers small, in clusters at the tip of branchlets. Fruit fleshy, dark, subglobose, and about 6 mm in diameter. Flowers and fruit can be found throughout most of the year.

Distribution: *North Island. Endemic.* Small populations of *Pseudopanax gilliesii* are scattered along the northern side of Whangaroa Harbour in North Auckland. The species has been reported from Little Barrier Island but this record has not been confirmed recently.

Habitat: Colonies of *Pseudopanax gilliesii* grow in coastal forest or scrub, often under regenerating kauri or puriri and taraire.

Conservation: Precise numbers remaining in the wild are not known, but only several hundred plants have been found. Although the species is probably not severely threatened by habitat changes at the moment, its limited geographic range and the small size of the known populations renders the species vulnerable to extinction in the wild. One population of about 20 plants occurs in a scenic reserve. *Pseudopanax gilliesii* is an attractive plant, occasionally found in cultivation. It strikes easily from hardwood cuttings and should grow readily from seed.

D. Given

Comments: In the past *Pseudopanax gilliesii* has sometimes been regarded as a hybrid because of its curious combination of divided and undivided leaves. *Pseudopanax*: a false *Panax; gilliesii*: after Thomas Gillies (1828-1889), a lawyer, M.P., district judge, and naturalist who founded the Auckland Institute together with Frederick Hutton and James Crawford, and explored the beaches of North Auckland with Thomas Kirk.

Status: Endangered.

middle-sized shrubs
(1-2m high)

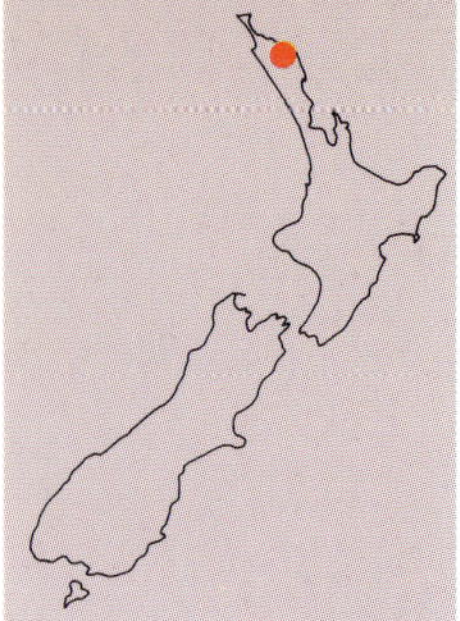

Hebe acutiflora

Scrophulariaceae

Description: Low, scrambling, sparingly branched shrub. Leaves linear-lanceolate, 2.5-9.5 cm x 3-9 mm, midribs yellowish orange. Leaf buds have no gap or sinus between the bases of the leaves. Flowers white to light mauve, 4-6 mm long, in elongate inflorescences up to 12 cm long. Seed capsules ovoid, about 3 x 2.5 mm. Flowers: January-April; Fruit: May-July.

Distribution: *North Island. Endemic.* For over one century *Hebe acutiflora* was known from a single location alongside the Kerikeri River in North Auckland. Although plants can still be found at this site, other populations have been discovered recently near Puketi, also in North Auckland.

Habitat: At both localities plants occur amongst scrub growing in damp soil alongside rivers and streams flowing through kauri forest.

Conservation: At least several hundred plants remain in the wild although numbers have not been accurately counted. One population occurs within a scenic reserve and the others are in a state forest. Mist flower appears to be competing with *Hebe acutiflora* at one site. *H. acutiflora* is an attractive plant and grows readily from cuttings, but few plants are in cultivation at present.

Comments: The exact relationship between this species and similar northern North Island species has yet to be investigated. Plants resembling *Hebe acutiflora* have been collected from the Bay of Islands, Herekino, and near the Wairoa River. *Hebe*: the Greek goddess of youth; *acutiflora*: with pointed flowers.

Status: Vulnerable.

S. Anderson

Hebe armstrongii

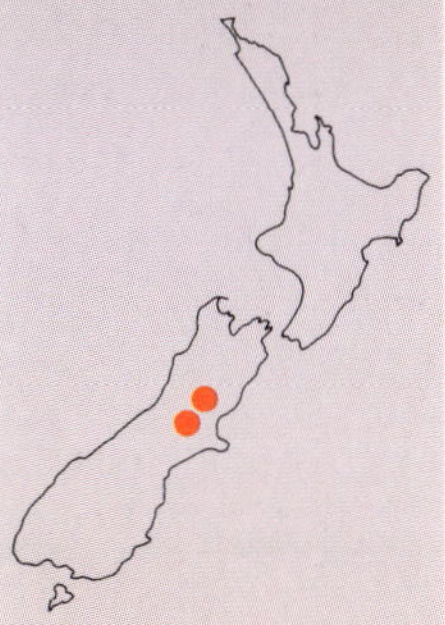

Scrophulariaceae

Description: Rounded, much-branched, yellowish green, whipcord shrub up to 1 m high. Branchlets cylindrical, 1.5-2 mm in diameter, arising from one side of the branch only. Pressed close to the branchlets are small scale-like leaves about 1 mm long. Their margins are yellowish and fringed with tiny hairs. Flowers in small clusters of 2-5 near branchlet tips, white to pale mauve, about 3 mm long. Seed capsules 3-3.5 x 1.5-2 mm. Flowers: December-January; Fruit: March-April.

Distribution: *South Island. Endemic. Hebe armstrongii* has been found only in Canterbury. It was discovered near the junction of the Lawrence and Clyde Rivers in the head of the Rangitata Valley in 1869 and specimens were later collected from the MacKenzie basin. Wild plants are now known from only two localities - near the Esk River and in the Castle Hill basin.

Habitat: This whipcord *Hebe* grows amongst bog pine on glacial outwash terraces covered in scrub and short tussock grassland.

Conservation: Wild populations are highly vulnerable. The Castle Hill population is within a scientific reserve but is extremely small and may be under long-term threat from intensive land-use close by. A recovery operation, begun in 1978, has involved nursery propagation of young plants and re-establishment in the reserve. Other sites are not protected. White-flowered forms of *Hebe armstrongii* are extensively cultivated throughout New Zealand. Plants grow readily from seed and cuttings are self-fertile, but older plants are prone to sudden die-back in summer.

Comments: When not in flower this species is almost indistinguishable from bog pine which often grows alongside *Hebe armstrongii*. *H. armstrongii* is distinguished from all other whipcord *Hebe*, except *H. cupressoides*, by its mauve flowers. *Hebe*: the Greek goddess of youth; *armstrongii*: after John Armstrong (1820-1902), who, together with W. Grey, first discovered this plant.

Status: Vulnerable.

B. Molloy

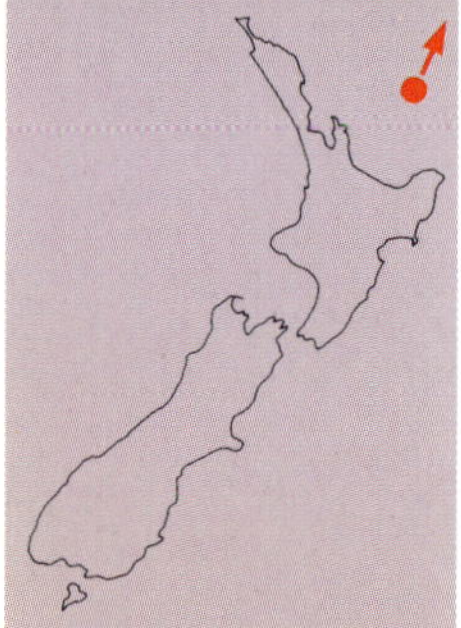

Hebe breviracemosa

Scrophulariaceae

Description: Laxly branched shrub up to 2 m, with green or yellowish branchlets covered in minute hairs. Leaves elliptic to oblong-lanceolate, 5-11 x 1.5-3 cm. The small white to pale lilac flowers, each about 6 mm in diameter, are in crowded inflorescences up to 10 cm long. Seed capsules are about 5 x 2.5 mm. Flowering and fruiting periods not known.

Distribution: *Kermadec Islands. Endemic*. This *Hebe* once flourished on Raoul Island, but now only one plant is known in the wild.

Habitat: *Hebe breviracemosa* appears to be a plant of partially open, tuff-covered ridge tops. At one time it was found on talus but appears to have been eliminated from these sites by goats.

Conservation: Until a few years ago *Hebe breviracemosa* was thought to have been totally extirpated through browsing by feral goats released on Raoul Island during the nineteenth century - the most recent collections having been made in 1908. Then, in February 1983, one young shrub was discovered on Raoul Island. It is hoped that *H. breviracemosa* will become re-established on Raoul following the eradication of the last goat on the island in November 1984. Cuttings and seedlings are now under cultivation and are easy to maintain.

Comments: *Hebe*: the Greek goddess of youth; *breviracemosa*: having flowers in short clusters.

Status: Endangered.

R. Lamberts

Hebe cupressoides

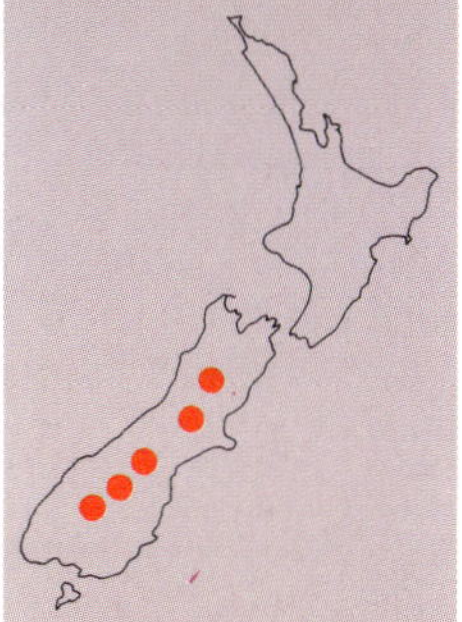

Scrophulariaceae

Description: Greyish green, much-branched shrub forming a symmetrical rounded bush up to 1-2 m, branches erect, slender and flexible, the smallest branchlets only 1-2 mm in diameter. Leaf pairs are widely spaced along the branchlets. Each fleshy, triangular, scale-like leaf is 1-1.5 mm long and has a characteristic turpentine smell when crushed. Flowers in clusters of 6-8 near the tips of the branchlets, small, white or purplish. Seed capsules narrowly oblong to linear-obovoid, about 2 x 1 mm.
Flowers: November-January; Fruit: March-May.

Distribution: *South Island. Endemic.* This *Hebe* once grew at about one dozen localities from Marlborough to Otago. The only records in recent years are from the Broken River basin, the Boyle River near Lewis Pass, the MacKenzie Basin and Ahuriri River catchment, and a tributary of the Shotover River in Otago. Other local populations may still exist in north Otago.

Habitat: *Hebe cupressoides* is a whipcord species found amongst sub-alpine to low alpine scrub and in tussock grassland on stony, well-drained sites. It is locally abundant on gravel riverbeds and terrace edges.

Conservation: Grazing by stock and burning are probably the main factors which have contributed to the decline in the total number of wild plants to an estimated several hundred specimens. A few plants are in a scenic reserve in the Broken River basin. The establishment of protected areas at two further sites, the only known populations containing numerous plants, is being negotiated. *Hebe cupressoides* is commonly cultivated in public and private gardens and is available commercially. It is relatively easy to grow from seed and cuttings.

Comments: *Hebe cupressoides* is distinguished from all other whipcord *Hebe* by its greyish green colour, very slender branches, densely branched habit, and widely spaced leaf pairs. *Hebe*: the Greek goddess of youth; *cupressoides:* resembling Cypress.

Status: Vulnerable.

C. Ogle

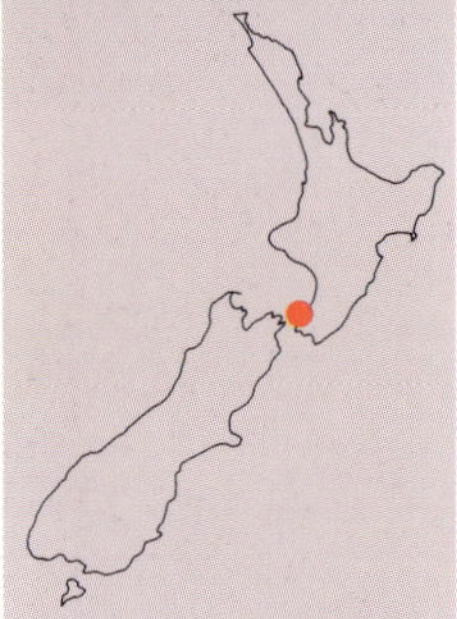

A. Robinson

Hebe elliptica var. crassifolia

Scrophulariaceae

Description: Much-branched shrub up to 2 m or more, branchlets very stout, pale green, predominantly hairless but may have 2 opposite rows of short hairs where the leaves are very close. Leaves pale green, almost fleshy, 1.5-3.5 x 0.7-1.5 cm, oblong to broadly obovate, fringed with fine white hairs along margins, a small, folded point at the tip. Leaf buds have a distinctive sinus between the bases of leaves. Flowers in inflorescences up to 6 cm, near branchlet tips, pure white, up to 1.2 cm in diameter. Fruit a sharply pointed, ovoid capsule, 6-10 x 4-6 mm.
Flowering and fruiting periods uncertain.

Distribution: *North Island. Endemic. Hebe elliptica* var. *crassifolia* is known from only two localities along the south-west coast of the North Island - near Titahi Bay and on Kapiti Island.

Habitat: All plants found in the wild so far grow on coastal cliffs and rocky coastal slopes.

Conservation: The plants near Titahi Bay are within a recreation reserve and Kapiti Island is a nature reserve. *Hebe elliptica* var. *crassifolia* is easily grown from cuttings and a few specimens are in cultivation.

Comments: The more widespread *Hebe elliptica* var. *elliptica* is similar to var. *crassifolia*, but has darker branchlets which are often densely covered with short, pale hairs, and white or bluish flowers. *Hebe*: the Greek goddess of youth; *elliptica*: shaped like an ellipse, referring to the leaves of *Hebe elliptica* var. *elliptica; crassifolia*: thick-leaved.

Status: Vulnerable.

Hebe speciosa

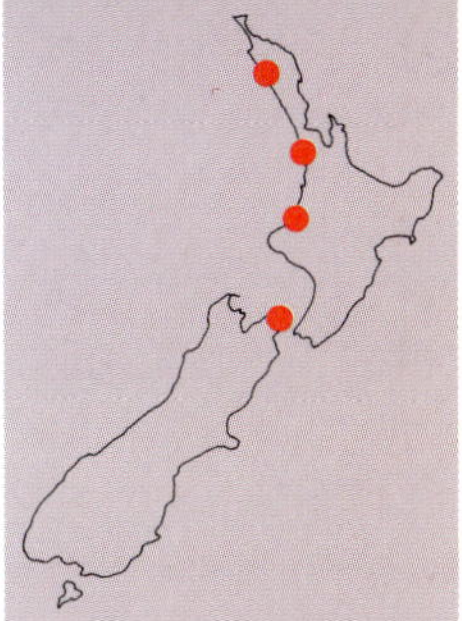

titirangi, napuka

Scrophulariaceae

Description: Rounded, bushy shrub up to 2 m, branchlets stout and slightly flattened. Leaves glossy, dark green, leathery to almost fleshy, broadly elliptic to obovate-oblong and 3.5-13 x 2.5-5.5 cm. Young leaves have purple margins and small hairs along the margins and midrib of their upper surfaces. Leaf buds have a distinctive sinus gap between the leaf bases. Flowers numerous, pale- to reddish magenta, about 7 mm long, in dense inflorescences up to 11 x 3 cm wide. Fruit an ovoid capsule, 4-6 x 2-3 mm.
Flowers: January-October; Fruit: seed capsules can be found throughout most of the year.

Distribution: *North Island, South Island, Endemic.* Several hundred plants grow near Maunganui Bluff, in North Auckland, and small populations have also been found south of Port Waikato in South Auckland and in north Taranaki. The only known South Island population grows in the Marlborough Sounds. Other colonies once occurred in the North Island near Hokianga Head, North Cape, and Wellington, but these have now disappeared.

Habitat: All the remaining populations grow on sea cliffs and steep slopes amongst low scrub or in relatively open situations.

Conservation: Numbers of *Hebe speciosa* have probably been reduced through browsing by livestock and erosion of the cliffs on which it grows. One colony is within a scenic reserve. Although *H. speciosa* is rare in the wild, plants which are usually of unknown origin are widely grown in public and private gardens. The species is a parent of many of the hybrid koromikos found in cultivation. It is easy to propagate from cuttings but is susceptible to hard frosts.

Comments: *Hebe speciosa* is similar to *H. macrocarpa* var. *brevifolia*, but is a larger plant with much longer flower spikes and wider leaves. *Hebe*: the Greek goddess of youth; *speciosa*: beautiful.

Status: Vulnerable.

D. Given

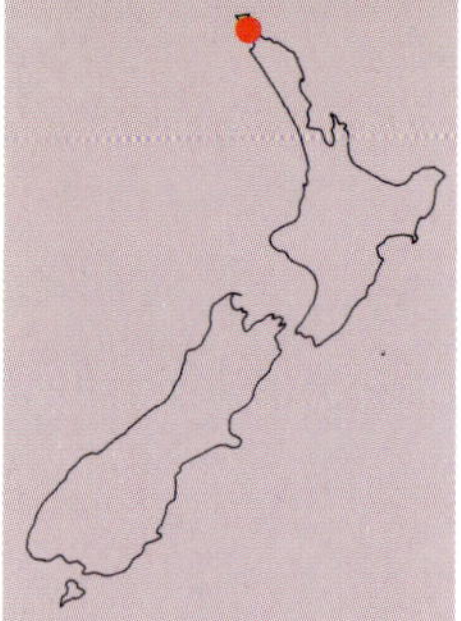

Hibiscus diversifolius

Malvaceae

Description: Stout sub-shrub up to 2 m with prickly stems which are often woody at the base. Leaves have bristly, entire or 3-5 lobed suborbicular to ovate laminae, 2-10 x 1-8 cm wide, margins irregularly toothed, petioles prickly, up to 9 cm long. Flowers are attractive, on the upper parts of the stems, pale to lemon-yellow with purplish black centres, 5-9 cm in diameter. Pale brown seeds are contained in an ovoid capsule about 2 cm long and enclosed within the bristly calyx.
Flowers and fruit can be found throught most of the year.

Distribution: *North Island, tropical Pacific region.* *Hibiscus diversifolius* occurs widely but sporadically in the tropical Pacific region. In New Zealand it grows only in the North Auckland region from North Cape to Whangaroa Bay.

Habitat: This plant occurs in coastal sites where there is virtually no frost, on seepages and boggy sites, often along the inland edge of sandy beaches.

Conservation: Approximately 12 wild populations of *Hibiscus diversifolius* are known. The species can be grown from seed (although results are variable) or cuttings, but is very susceptible to frost. It is occasionally seen in cultivation.

Comments: *Hibiscus diversifolius* includes a variety of forms which are not well understood. New Zealand wild plants differ from many plants in cultivation under this name, even in New Zealand. *Hibiscus trionum*, a species of probably adventive origin, is also found in coastal sites in North Auckland. It is a much smaller herbaceous plant (up to 60 cm high) with more saucer-shaped flowers and bristly rather than prickly stems, and is annual or biennial.
Hibiscus: the old Latin name for marshmallow; *diversifolius*: with leaves of more than one shape.

Status: Vulnerable.

D. Given

Olearia semidentata

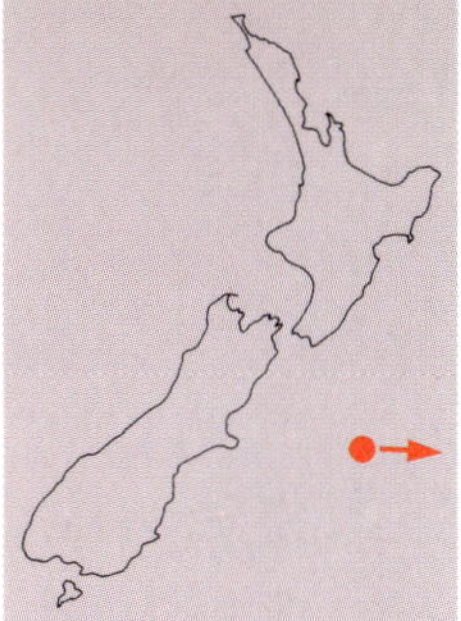

makora; Chatham Island aster

Asteraceae

Description: Rounded shrub usually up to 2 m but is sometimes taller, slender branchlets clad in soft white hairs. Leaves narrow lanceolate with widely spaced teeth towards the tip, covered with dense white hairs on the lower surface, laminae 2.5-7 cm x 4-17 mm. Flower heads on slender, hairy stems 3-8 cm long, most numerous towards branchlet tips. Flower heads (capitula) up to 3.5 cm in diameter, usually purple but fading to pink with age. Fruit on achenes about 3 mm long, crowned by hair-like bristles up to 7 mm long. Flowers: November-March; Fruit: January-June.

Distribution: *Chatham Islands. Endemic. Olearia semidentata* is found on Chatham Island and has also been recorded from South East Island. On Chatham Island it is locally abundant on open parts of the Southern Tablelands. It also occurs at the northern end of the island.

Habitat: Chatham Island aster frequently grows in association with *Sporadanthus traversii* in restiad-scrubland on wet, peaty soils.

Conservation: Although *Olearia semidentata* can withstand some burning and rapidly recolonise burnt areas, burning accompanied by drying out of the soil and cattle trampling has probably eliminated this species from some areas. Continued efforts to use restiad-scrubland for farming and the possible development of mining for peat wax and fuel oil threatens this species on Chatham Island. Plants occur in Rangaika and Tuku Reserves but adequate conservation of other populations is still needed. This spectacular plant is occasionally cultivated both in New Zealand and overseas. It can be readily propagated from seed and cuttings, but cultivated plants frequently succumb to root pathogens.

Comments: *Olearia semidentata* is very similar to *O. chathamica* in appearance, but it is a smaller plant with somewhat narrower, more erect and less glossy leaves. *Olearia:* after Johann Gottfried Oelschlaeger (1635-1711), *semidentata:* partially toothed.

Status: Vulnerable.

D. Given

Pittosporum pimeleoides

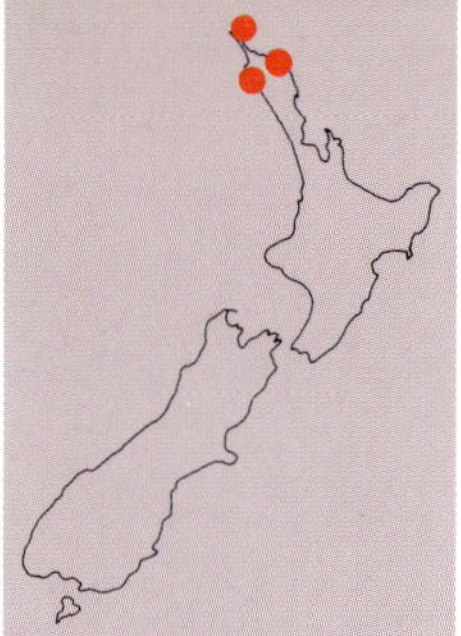

Pittosporaceae

Description: Low shrub up to 2 m tall with very slender, flexuous branches and main stem which may descend to the ground and take root. Leaves alternate, in whorls of 2-5 or crowded towards branchlet tips, up to 5 cm x 7 mm, narrowly elliptic, oblanceolate, linear-oblong or linear, tip with short, slender point. Male and female flowers are borne on separate plants. Male flowers in small clusters of 5-10, female flowers solitary or in groups of 2-3, petals 7-12 mm long, pale yellow with a central crimson streak. Black seeds are in a 2-valved ovoid capsule, 5-15 mm long with a pointed, beak-like apex. Capsules covered in silky hairs when young, woody and dark brown with age.
Flowers: April-July; Fruit: usually reaches maturity in the following year.

Distribution: *North Island. Endemic*. In the past *Pittosporum pimeleoides* has been recorded from many lowland and coastal forest localities in the North Auckland region, but it is now known from less than 12 sites, most of which are between North Cape and Whangarei.

Habitat: *Pittosporum pimeleoides* usually grows in semi-open sites on ridges or well drained slopes. It is often associated with kauri, toatoa, mingimingi, and mairehau, but is also found in tea-tree scrub.

Conservation: Many of the sites from which this species was once recorded have been cleared for farming or forestry and this practice now threatens some surviving populations. Several hundred plants are known in the wild. The species occurs in two scenic reserves in the Bay of Islands, an ecological area within a state forest, and two proposed reserves. A few plants are in cultivation in botanic and private gardens. It can be easily grown from cuttings or seed.

E. Cameron

Comments: *Pittosporum pimeleoides* is closely related to *P. michiei* (which is sometimes included in *P. pimeleoides* as variety or subspecies *maius* or *major*), but can be distinguished by the size and shape of the leaves and its erect or inclined rather than prostrate stems. *P. pimeleoides* also resembles mingimingi, mairehau, and juvenile kanuka, but unlike *P. pimeleoides*, mairehau has very aromatic leaves with bluntly pointed tips, mingimingi leaves have distinctive parallel veins, and kanuka leaves are alternate and aromatic when crushed. *Pittosporum*: pitch seed, referring to the resinous pulp surrounding the seeds; *pimeleoides*: resembling *Pimelea*.

Status: Vulnerable.

Pomaderris oraria var. novae-zelandiae

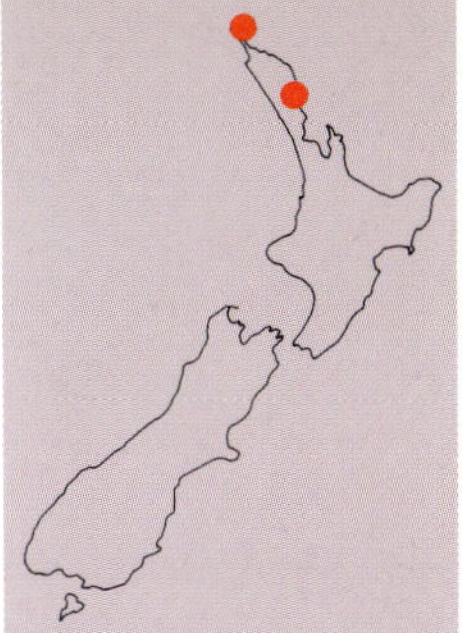

Rhamnaceae

Description: A scrambling to erect shrub up to 1.5 m tall. Adult leaves oblong 9-50 x 6-20 mm, margins slightly toothed and not recurved, lower surface covered with star-like clusters of both white and brown hairs. Juvenile leaves are very variable, some are less than 6 mm long, conspicuously toothed, and covered on both surfaces in a dense layer of short hairs. Flowers greenish, lacking petals, about 4 mm in diameter, borne in elongated, leafless panicles. Fruit a ripe capsule half-immersed in the calyx-tube which retains its sepals and bears hairs similar to those on the lower surface of leaves. Three cocci within the capsule each contain a single, dark brown seed, 1-2 x 1 mm.
Flowers: October-November; Fruit: December-February.

Distribution: *North Island. Endemic.* A large population of several hundred plants grows near North Cape and a few smaller populations are found near Whangarei Harbour.

Habitat: *Pomaderris oraria* var. *novae-zelandiae* grows in low coastal scrub on hillsides, cliffs, and rock outcrops.

Conservation: One of the sites where *Pomaderris oraria* var. *novae- zelandiae* survives is within a scientific reserve and another is in a scenic reserve. As this species is usually a component of successional scrub communities, the sites where it is reserved may require interventionist management to maintain them in a scrub state. A few plants are grown in scientific and private gardens. It is hardy in cultivation and grows readily from seed and cuttings.

Comments: This species is similar to *Pomaderris rugosa*, but that is a slightly taller shrub which grows up to 3 m and has entire leaf margins, more compact clusters of cream flowers, and does not retain the sepals on the capsule. *Pomaderris*: from *poma*, lid, and *derris*, skin, referring to the membranous lid on the inner wall of each coccus which opens to release the seed; *oraria*: coastal; *novae-zelandiae*: of New Zealand.

Status: Endangered.

E. Cameron

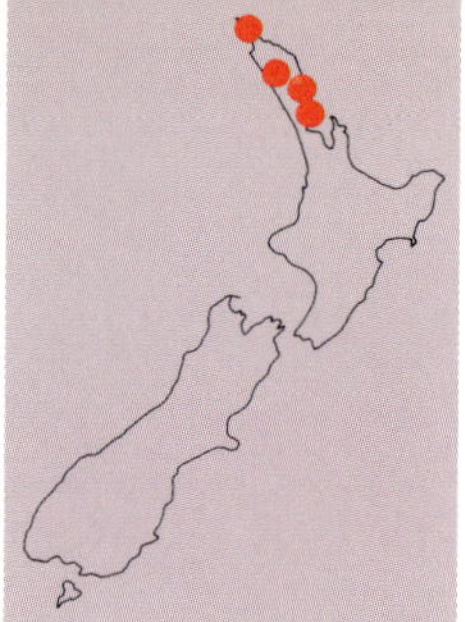

Pomaderris phylicifolia var. polifolia

Rhamnaceae

Description: A low bushy shrub up to 1.5 m tall. Adult leaves, with very short petioles, laminae narrowly elliptic to narrowly oblong, 5-15 x 2-5 mm, margins entire and recurved, upper surfaces are clad in soft, erect hairs, lower surfaces with dense star-shaped clusters of white hairs. Juvenile leaves 8-30 x 5-10 mm and lacking recurved margins. Flowers cream, lacking petals, 4-5 mm in diameter, grouped in small clusters at branchlet tips and in the axils of the leaves. Fruit a ripe capsule 4 mm long and one-third immersed in the calyx-tube which is clad in simple white hairs and does not retain its sepals. Three cocci within the capsule each contain a single, shiny, dark brown seed up to 2.5 x 1.5 mm. Flowers: October-November; Fruit: November-December.

Distribution: *North Island. Also found in Australia. Pomaderris phylicifolia* var. *polifolia* is known from about 10 sites between North Cape and Houhora Harbour and 4 locations from Waipoua to Silverdale. *P. phylicifolia* var. *polifolia* is also found in Australia but has yet to be fully compared with its New Zealand counterpart.

Habitat: This plant usually grows amongst stunted manuka and sedges on infertile gumlands.

Conservation: The largest known population of *Pomaderris phylicifolia* var. *polifolia* grows in a farm park and on adjoining land in private ownership. The colony near Silverdale has a fungal disease and appears to be dying. *P. phylicifolia* var. *polifolia* is readily grown from seed, but cuttings can be difficult to root. It makes an attractive garden plant and is increasing in popularity.

O. John

Comments: Specimens with a hairless upper leaf surface were once found near Te Kao, but had disappeared from the wild by early 1979. *Pomaderris ericifolia* is a similar but smaller shrub up to 1 m high with smaller leaves 2-8 x 1-2 mm. It has star-shaped clusters of hairs on the calyx-tube and lacks petal-like outgrowths from the filaments common in flowers of cultivated specimens of *P. phylicifolia* var. *polifolia*. *Pomaderris*: from *poma*, lid, and *derris*, skin, referring to the membranous lid on the inner wall of each coccus which opens to release the seed; *phylicifolia*: having leaves like those of *Phylica; polifolia:* having leaves like those of germander (*Teucrium polium*).

Status: Vulnerable.

low and prostrate shrubs

(less than 1m high)

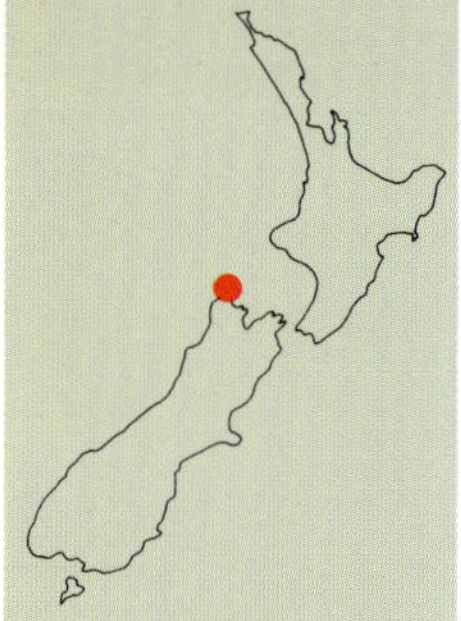

Carmichaelia fieldii

Fabaceae

Description: Slender, prostrate shrub with branches up to 1 m long, branchlets narrow compressed and grooved, 1-2 mm wide, and sparsely covered in soft hairs. Leaves short-lived, only present on young plants and actively growing branchlets, composed of 1-7 linear leaflets. Flowers small, white with purple veins, about 4 mm long. Seed pods dark brown to black, 3-4 mm long, ovoid with a stout to slender beak about 1 mm long. Each pod contains 1 or 2 yellowish brown seeds with or without black markings.
Flowers: November-February; Fruit: February-April.

Distribution: *South Island. Endemic. Carmichaelia fieldii* was first discovered early this century on a small island in West Whanganui or Westhaven Inlet, in north-west Nelson, but has not been seen here since. The species was rediscovered at two sites each several kilometres to the north and south of the inlet in the late 1970s.

Habitat: Specimens have been collected from wind-swept sandstone and conglomerate ledges near the sea and from coastal herbfield.

Conservation: This broom could be readily overwhelmed by competing plants and browsing animals. One of the two known populations has been fenced off; numbers here are not known. The other occurs in a farm park and consists of about 10 plants. A few *Carmichaelia fieldii* are in cultivation; it can be easily grown from seed or cuttings.

Comments: The precise taxonomic status of closely related *Carmichaelia floribunda, C. lacustris,* and *C. prona* is uncertain and they may be synonymous with *C. fieldii. Carmichaelia floribunda* was collected once by G. Simpson from Awatere Valley, Marlborough; *C. lacustris* was collected only from the shores of Lakes Manapouri and Te Anau but has not been seen in recent years; and, *C. prona* was recorded from margins of small lakes between Cass and Porters Pass in Canterbury, where it is now presumed to be extinct. *C. prona* is known in cultivation. *Carmichaelia*: after Dugald Carmichael (1772-1827), a Scottish army officer who was stationed in South Africa. He wrote an account of Tristan da Cunha; *fieldii*: after William Field (1861-1944), M.P., and amateur botanist who discovered this species with Bernard Aston.

Status: Endangered.

D. Given

Myoporum debile

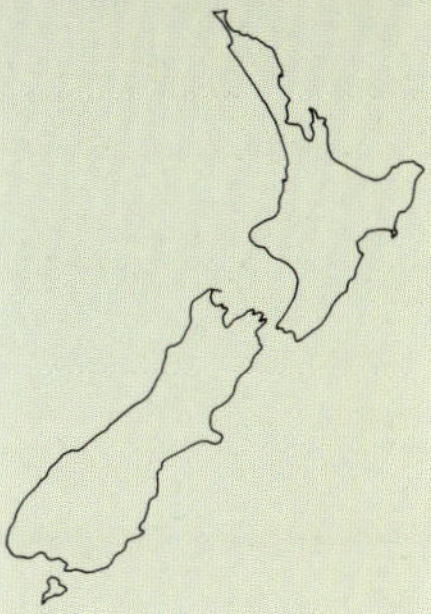

Myoporaceae

Description: Low, sprawling shrub with slender branches up to 50 cm long, branchlets are often reddish on the upper surface and covered in small tubercles. Leaves thin, linear-lanceolate, 4-10 cm x 5-10 mm, margins usually with small teeth near the tip and the base of the leaf, but occasionally entire or fully toothed. Numerous, rather obscure, semi-transparent spots or oil glands can be seen near the leaf margins. Flowers white to purplish, 6-8 mm long, solitary or in pairs in the leaf axils. Globose drupes are white below, reddish purple above, approximately 9 x 10 mm. Flowers: January-May; Fruit: May-September.

Distribution: *North Island. Also in Australia. Myoporum debile* was once found at two localities along the west coast of the North Island near Kawhia Harbour and Raglan Harbour, but both of these populations are now thought to be extinct.

Habitat: This species grew along rock ledges of steep terraces and cliffs.

Conservation: The colony near Kawhia Harbour disappeared in 1981 as a result of denudation by feral horses, sheep, and goats, and increased erosion caused by poorly situated drains from local baches. Parts of the coastline near Raglan Harbour have been burnt in recent years and it is thought that the second colony has also been destroyed as a result. *Myoporum debile* is grown as a garden plant in England, Australia, and New Zealand. It is easily propagated from cuttings and seed, but is damaged by frost.

Comments: The ngaio (*Myoporum laetum*), which grows as a shrub or small tree up to 10 m high, is related to *Myoporum debile*, but its growth form contrasts sharply with the low sprawling habit of *M. debile. Myoporum*: closed pores, referring to the glands in the leaves; *debile*: weak.

Status: Presumed extinct.

D. Given

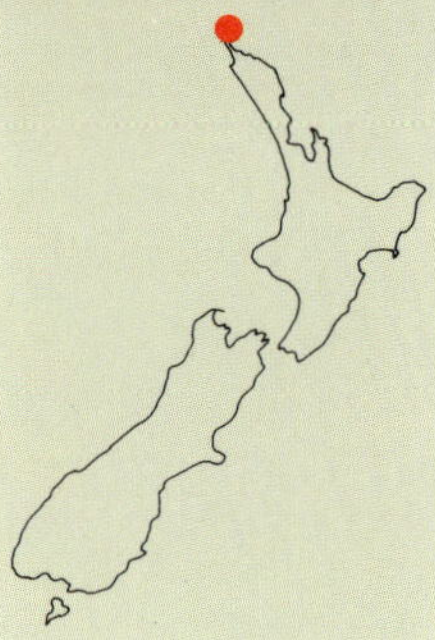

Pittosporum michiei

Pittosporaceae

Description: Slender, prostrate shrub producing roots at intervals along long, lax branches, up to 4 m long and about 5 mm in diameter. Leaves stiff, usually arranged along the stem in whorls of 2-5, obovate to elliptic-obovate or sometimes elliptic, usually 10-25 x 4-12 mm, occasionally up to 40 x 25 mm. Flowers fragrant, about 11 mm long, petals yellow with a central crimson streak. Male and female flowers on different plants; male flowers most often grouped in clusters of 5-6, female flowers usually solitary. The black seeds develop in an ovoid 2-valved capsule which is 6-12 mm long including the prominent "beak". Each capsule is covered in silky hairs when young, becoming woody and dark brown with age.
Flowers: July-August; Fruit: seed capsules reach maturity in the following year.

Distribution: *North Island. Endemic.* Definite *Pittosporum michiei* is known only from North Cape. Another record from Cape Maria van Dieman has not been recently confirmed.

Habitat: This plant is restricted to areas of scrub and low open forest on steep serpentine sea cliffs. It may once have been more abundant in low scrub on a level plateau behind the cliffs.

Conservation: In recent years it has been estimated that up to 1000 specimens remain in the wild. Burning of the plateau scrub before 1967 may have eliminated plants growing on the plateau behind the cliffs, but fortunataly part of the cliffs and the plateau is now protected as a scientific reserve. Serpentine rock used for fertiliser has been quarried from a site adjacent to the reserve. The species is easy to grow, flowers well, and is readily propagated from layers and cuttings.

B. Given

Comments: This is one of about 15 plant taxa (some of which are undescribed) which may be virtually or entirely restricted to the North Cape area. *Pittosporum michiei* is closely related to *P. pimeleoides* and is sometimes given varietal or subspecific status under *P. pimeleoides* (as var. or subsp. *maius* or *major*). *Pittosporum*: tarry or sticky seed, referring to the resinous pulp surrounding the seeds; *michiei*: Ross Michie (1894-1987) of Kaitaia who discovered numerous colonies of rare plants in the Auckland region.

Status: Vulnerable.

climbers and scramblers

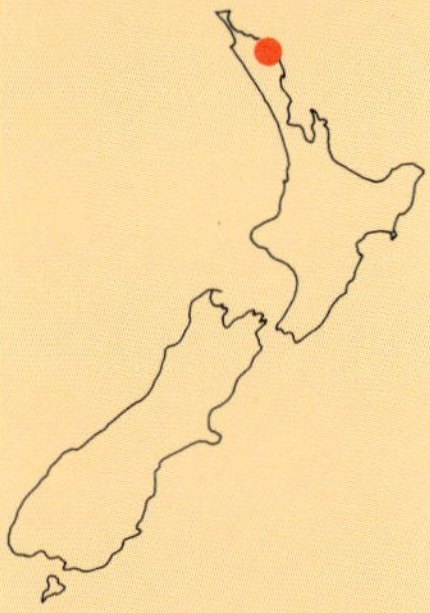

Calystegia marginata

Convolvulaceae

Description: Climbing plant, stems slender, twining up to 2 m long, rhizome creeping and branching. Leaves with petioles up to 6 cm, laminae with sharply pointed tips and sagittate bases, 2.5-10 x 1.5-4.5 cm, narrower and more pointed than the laminae of other New Zealand species of *Calystegia*. Flower stems have narrow wings which terminate at 2 broadly ovate bracts enclosing the calyx at the base of each flower. Flowers white, about 12 mm in diameter. Fruit a globose capsule which usually contains 4 seeds.
Flowers: November-March; Fruit: January-May.

Distribution: *North Island. Also found in eastern Australia*. Sparse populations of *Calystegia marginata* occur near Cape Brett and in the Whangaruru-Waikare Inlet area of the Bay of Islands; a few plants have also been discovered on three islands nearby. Another population was recently discovered further south near Hukerenui. Other North Auckland populations were once found near Kaingaroa, Fairburn, Kaiaka, Kaikohe, Paihia, Maungatapere, and Kopehua but these have now disappeared.

Habitat: *Calystegia marginata* has been found along bush margins, in open, tea-tree scrub and amongst low shrubs, blackberry and other weeds.

Conservation: This species occurs locally within two scenic reserves and populations have also been reported from Russell State Forest. Some of the plants at one site were killed during spraying for blackberry and this population is now almost extinct. *Calystegia marginata* is rarely known in cultivation.

Comments: *Calystegia*: covered calyx, referring to the bracts which enclose the base of each flower; *marginata*: edged with a distinct margin, referring to the winged flower-stems.

Status: Vulnerable.

S. Anderson

Carmichaelia kirkii

Fabaceae

Description: Semi-liane or scrambling shrub, stems usually supported by other plants or interlaced amongst themselves, branchlets grooved, 1-2 mm wide and sometimes sparsely hairy. Leaves are up to 25 x 8 mm, and consist of 3-5 inversely heart-shaped leaflets, each up to 7 x 5 mm. Flowers in clusters of 2-6, 6-10 mm long, white to cream with purple veins and blotches. The dark brown seed pods remain on the plant for many months. They are 8-15 mm long and have a stout beak at the tip, 4-5 mm long, each pod with 2-4 pale grey to pale mauve seeds with distinctive black markings.
Flowers: November-December; Fruit: January onwards.

Distribution: *South Island. Endemic*. Populations of *Carmichaelia kirkii* have been reported from New Brighton, Christchurch, and the Rakaia Gorge in Canterbury, and from the Otepopo River, Gimmerburn, and Sowburn in Otago, but all these populations may now be extinct. The present distribution of *C. kirkii* is very discontinuous. The main populations grow on the Benmore Range in South Canterbury; there are smaller outlying populations elsewhere in the Mackenzie Basin, near Omarama and Tarras, and alongside some tributaries of the Cardrona and Lindis Rivers in Otago; and near Oxford in North Canterbury. A single plant was recently found in the valley of a tributary of the Awatere River in Marlborough.

Habitat: *Carmichaelia kirkii* has been found on scrub-covered mountain slopes and steep boulder fields, along streams and lake margins, and in swamps where the soil is very moist.

D. Given

Conservation: Conversion of this plant's habitat to farmland and browsing by livestock has resulted in decline, and only a few hundred plants may remain in the wild. One population was eliminated by the raising of Lake Pukaki for hydro-electricity generation. A colony of about 60 plants near Oxford in North Canterbury has been fenced and young plants established in an attempt to ensure this small population survives. Priority needs to be given to the establishment of reserves containing *Carmichaelia kirkii* at more southern sites. This broom is easily grown from seed and cuttings and is occasionally found in cultivation.

Comments: *Carmichaelia kirkii* was described from specimens collected in the Cardrona Valley in the late 1800s. It was rediscovered here early in 1985. *C. kirkii* is the only species in the genus which grows as a liane and is also one of the few species which can be quite leafy, although the leaves are shed in winter or in drier localities exposed to sun and wind. Some confusion surrounds the taxonomy of the species. Plants which are sometimes included in *C. kirkii* as var. *strigosa* are of uncertain taxonomic status, *C. gracilis* is now generally included in *C. kirkii*, but it has been suggested that Otago and Canterbury plants may be distinct.
Carmichaelia: after Dugald Carmichael (1772-1827), a Scottish army officer who was stationed in South Africa. He wrote an account of Tristan da Cunha; *kirkii*: after Thomas Kirk (1828-1898), an English botanist who arrived in New Zealand in 1863.

Status: Vulnerable.

Helichrysum dimorphum

Asteraceae

Description: Small shrub or scrambling liane, stems slender and flexible, up to several metres long. Sessile leaves are of 2 different forms which may alternate along a single stem. Widely spreading leaves of the first form 2-7 x 1-3 mm, usually elliptic to oblong or sometimes obovate with recurved margins, upper surfaces almost hairless, lower surfaces covered in dense hairs. Narrow, scale-like leaves of the second form 2-4.5 x about 1 mm, held close to the stem or pressed tightly against it, upper surfaces clad in dense, white hairs, lower or outer surfaces moderately to sparsely hairy. Flowers in small white heads (capitula) up to 5 mm in diameter, at the tips of branchlets. Fruit an achene 1.5-2 mm long, crowned by white bristles about 4 mm long.
Flowers: December-April; Fruit: February-May.

Distribution: *South Island. Endemic. Helichrysum dimorphum* grows in the middle part of the Waimakariri Valley and tributaries in Canterbury.

Habitat: Steep terrace faces and the sides of gorges are the usual habitat of this species. It is often associated with shrubs such as matagouri, *Corokia cotoneaster*, and *Coprosma* species.

Conservation: Although *Helichrysum dimorphum* has probably always been rather local in distribution, it was once more widespread in the central part of the Waimakariri catchment. Numbers have declined in recent years and only 200 to 300 plants are known to survive in the wild. A large colony of several hundred plants was destroyed by aerial spraying for noxious weeds in 1978 and some other populations have been overwhelmed by introduced weeds such as broom. None of the sites where *H. dimorphum* occurs are reserved, and only a few plants are known to be in cultivation. It can be grown from seed and from tip cuttings.

Comments: This is the only New Zealand member of the genus which has a lianoid growth form. The factors causing the two different leaf forms are not known. *Helichrysum*: sun gold; *dimorphum*: occurring in two different forms.

Status: Endangered.

D. Given

Tecomanthe speciosa

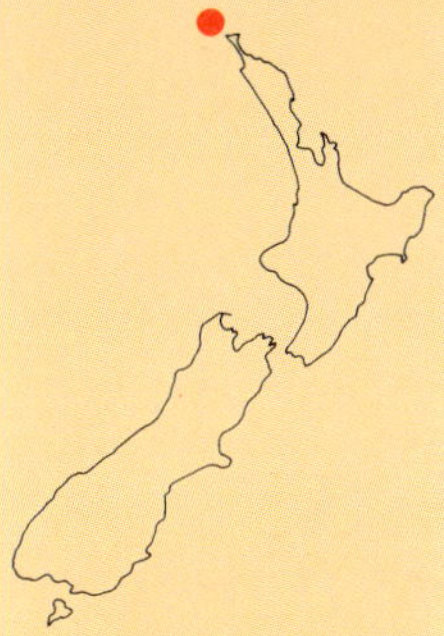

Bignoniaceae

Description: Woody liane with stout stems up to 10 m long. Leaves pinnate, of 3-5 glossy, dark green leaflets which are broadly elliptic to broadly obovate, 5-16 x 3-8 cm. Clusters of up to 50 flowers develop in the leaf axils or on old wood above the leaf scars. Individual flowers tubular or funnel-shaped, 6-8.5 cm long, petals creamy white and flushed with pale green, hairy on the outer surface. Fruit a cylindrical capsule up to 19 cm long containing many flattened seeds; each about 1 x 3 cm including the membranous wings.
Flowers: April-June; Fruit: seed capsules have never been found on wild plants; capsules on cultivated plants mature in February-May of the following year.

Distribution: *Three Kings Islands. Endemic. Tecomanthe speciosa* occurs only on Great Island in the Three Kings group, north of the New Zealand mainland.

Habitat: The oldest vine on Great Island has scrambled to the top of a young pigeonwood growing in shallow, damp soil overlying bedrock at an altitude of 150 m. The species' original habitat was probably mixed coastal or semi-coastal forest.

Conservation: When *Tecomanthe speciosa* was first discovered on Great Island in 1945 only one vine was found. Since then the main vine has produced another six rooted vines by layering. In 1982 no flowers or fruit were seen and no seedlings were found. Introduced goats are probably chiefly responsible for the rarity of this species, although forest clearing by the Maori who occupied the island for 2 centuries before 1840 may have also severely modified the plant's habitat. The goats were exterminated in 1946 and the Three Kings Islands are now a Nature Reserve. *Tecomanthe speciosa* has never been known to reproduce sexually in the wild although it is self-fertile. Despite its ability to reproduce asexually to a limited degree, the species' long term-survival in the wild is not certain. It is an attractive plant available commercially, and is increasing in popularity. It is fairly readily propagated from seed, layers, and cuttings, and grows well in the warmer parts of New Zealand and in cooler areas where it is protected from frosts.

Comments: *Tecomanthe speciosa* is New Zealand's only representative of this large tropical family. Its flowers have several features which appear to be adaptations for pollination by bats, but bats have never been found on the Three Kings Islands. *Tecomanthe*: having flowers like *Tecoma, speciosa*: beautiful.

Status: Endangered.

J. Braggins

herbaceous plants

Hydatella inconspicua

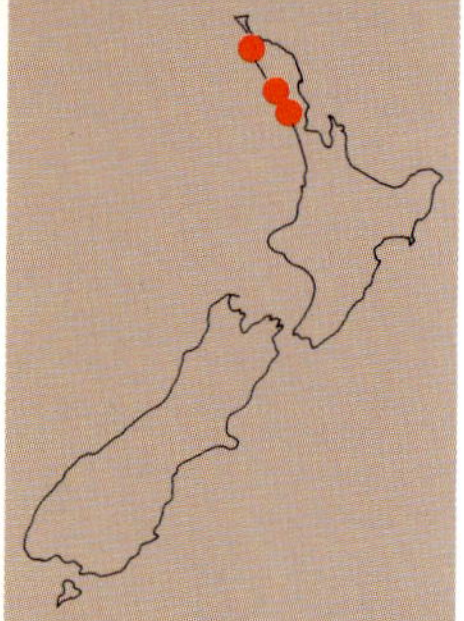

K. West

male flowers

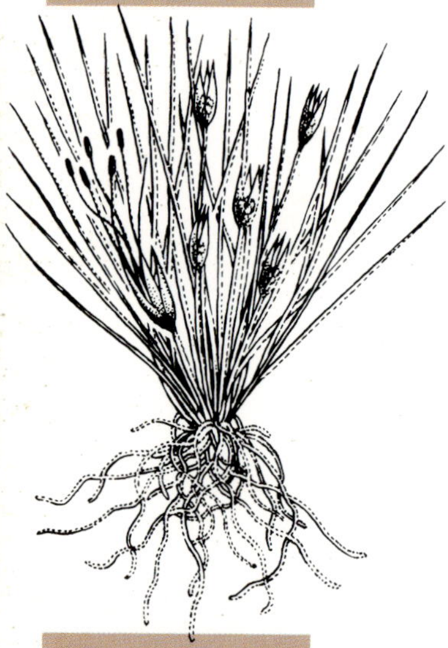

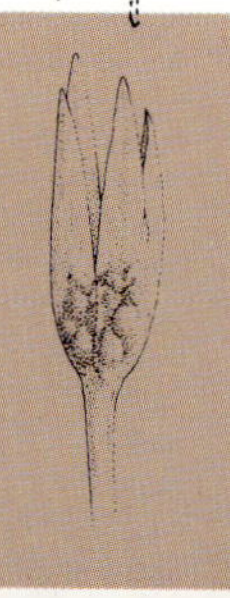

female flowers

Hydatellaceae

Description: Inconspicuous, grass-like herb forming erect tufts of sharply pointed, linear to thread-like leaves 2-4 cm x 0.5 mm. Leaves colourless in the lower part and bright green in the upper third, often becoming purplish red with a central red vein at maturity. Flower stems bright red, 2-3 mm long at flowering stage, later elongating to 3-4 cm when fruit is set, each stem usually terminating in a cluster of either 2-8 male flowers or about 10 female flowers (a few clusters of up to 24 female flowers or both male and female flowers have also been found). Individual flower clusters surrounded by 2-5 translucent, membranous bracts 3-5 mm long. Female flower of a rounded, stalked ovary with 5-10 styles, about 0.5 x 0.4 mm wide. Male flowers rarely found, bright red, the single stamen about 1 cm long at maturity. Fruit minute, reddish brown, ellipsoid-ovoid with a dark, pointed tip.
Flowers: October-January; Fruit: December-February.

Distribution: *North Island. Endemic*. This small aquatic plant is known from 13 small lakes scattered along the west coast of North Auckland.

Habitat: *Hydatella inconspicua* grows in shallow water on the shores of dune lakes in moderately wave-exposed sites where it may be submerged up to a depth of 1 m or rarely up to 2 m. Plants commonly grow in sand, but sometimes in silt and organic mud. It often occurs amongst open stands and in the shelter of tall emergent sedges.

Conservation: Lake shore pugging by cattle, increases in water nutrients, and extreme water level fluctuations are probably the main threats facing this plant. Exotic weeds such as *Egeria densa* and *Elodea canadensis* could also endanger its survival if they are introduced into the lakes in which *Hydatella inconspicua* occurs. There is an urgent need to study the effects of cattle grazing on lake shores and securely reserve some of the lakes where this species grows, as well as buffer zones around them. *H. inconspicua* has been successfully transferred from the wild and grown in tanks at Dargaville High School and outdoors at Ruakura Agricultural Research Centre. Most attempts to artificially germinate seed have been unsuccessful.

Comments: The genus *Hydatella* contains only five species. *H. inconspicua* is endemic to New Zealand, the remainder to Australia. Two of the Australian species have not been seen since the turn of the century. *Hydatella* and the related genus *Trithuria* were formerly considered members of the family Centrolepidaceae, but recent treatments place these small genera in the family Hydatellaceae, where they together constitute the order Hydatellales. *Hydatella*: living in water; *inconspicua*: inconspicuous, not readily noticed.

Status: Vulnerable.

Myriophyllum robustum

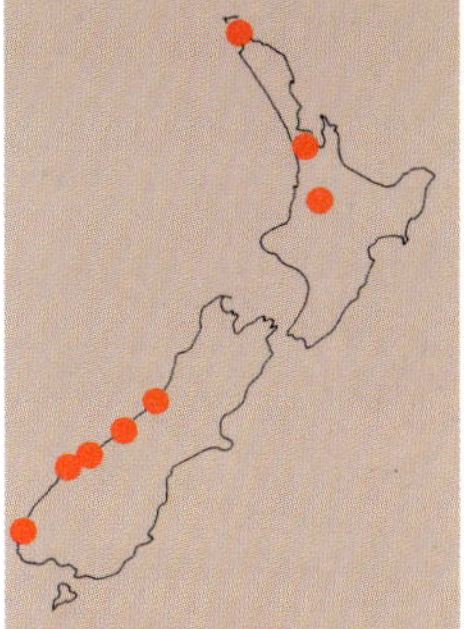

stout water-milfoil

Haloragaceae

Description: Stout aquatic herb with fleshy, sparingly branched stems up to 160 cm long, the tips of which emerge above the water surface. Submerged leaves often decay early, but when present are arranged in whorls of 5-7 ovate leaves with rounded tips, each leaf 3-5 cm x 6-8 mm, divided into 24-30 thread-like pinnae 15-25 mm long. Yellowish green to bluish green emergent leaves crowded towards the tips of the stems in whorls of up to 7, oblong with a sharply pointed tip, 15-35 x 4-10 mm wide, divided into 26-32 linear pinnae 2-7 mm long. Flowers tiny, solitary, green to pinkish with both stamens and carpels, borne in the axils of the upper emergent leaves. Fruit globose, 1.5-2 mm long, splitting into 4 mericarps.
Flowers: September-May; Fruit: November-April.

Distribution: *North Island, South Island. Endemic.* Past distribution records of stout water-milfoil include numerous lowland localities throughout the North Island and the north of the South Island, but it is now known from only 8 individual sites in the North Island and less than 15 along the west coast of the South Island.

Habitat: The preferred habitat is pools and slowly flowing streams in peaty lowland swamps but in the North Island it is now largely restricted to swampy kahikatea forest and shallow lake margins.

Conservation: In 1914 Thomas Cheeseman noted that *Myriophyllum robustum* covered large stretches of swamps but also observed that its distribution was being severely reduced by draining and burning of this habitat. Continued draining of wetland areas, forest clearing, open-cast mining, and invasion by weeds such as *Ludwigia, Alisma,* and *Salix* spp. has resulted in the extinction of this plant at many sites and its long-term survival in the North Island may be in doubt. A few populations occur within protected areas (two National Parks, a wildlife management reserve and a farm park); a conservation covenant for the protection of one North Island site is being negotiated. *M. robustum* can be easily grown from cuttings and fruit, and may make a good aquarium plant. Material from several colonies is in cultivation.

Comments: Although it is somewhat similar to the introduced *Myriophyllum aquaticum* used in aquaria, *M. robustum* can be distinguished by its pinkish rather than white flowers, sharply-pointed and spreading rather than erect and rounded emergent leaves, and the presence of fruit in the leaf axils. Male flowers and fruit have not been found on New Zealand specimens of *M. aquaticum*. Their larger emergent leaves distinguish both of these plants from other species of *Myriophyllum* found in New Zealand. *Myriophyllum*: myriad-leaved; *robustum*: strong, well built.

Status: Vulnerable.

E. Cameron

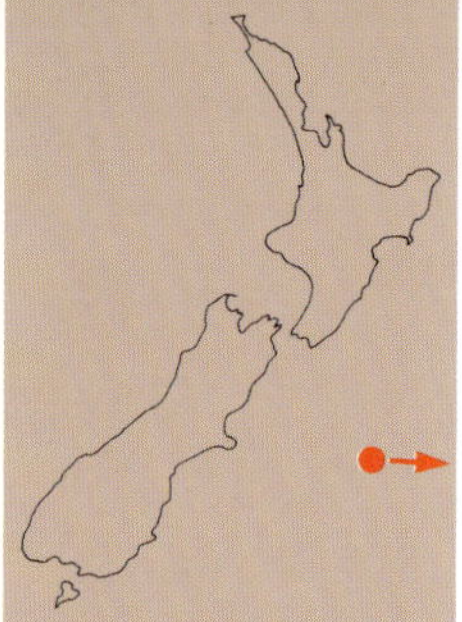

Embergeria grandifolia

Chatham Island sow thistle, great sow thistle

Asteraceae

Description: Robust, milky-sapped herb forming large clumps up to 1.5 m tall, spreading by stout, fleshy rhizomes. Leaves at base of the plant, oblong, up to 1 m x 20 cm, scabrid on upper surfaces, irregularly divided into sharply toothed lobes. The leaves borne higher up the stem are somewhat smaller. Flower heads (capitula) pale purple, up to 4 cm in diameter. Fruit of achenes about 4 mm long, with obvious longitudinal ribs and crowned by numerous white hair-like bristles. Flowers: December-March; Fruit: January-May.

Distribution: *Chatham Islands. Endemic. Embergeria grandifolia* is found on Chatham, Pitt, Mangere, and Little Mangere Islands and some other islets in the Chatham group. Early reports indicate that it was probably more widespread on these islands in the past.

Habitat: This thistle occurs sporadically on coastal dunes, in coastal forest and scrub on sandy soils, and to a lesser extent on cliffs and coastal talus. It may have been a major component of turf on stable sand dunes in many areas until it was eaten out by animals.

Conservation: Grazing by farm stock has severely depleted this species. Several thousand plants probably survive on the Chatham Islands but many populations consist of only a few plants. Some plants grow in a nature reserve on Mangere Island, but the best populations on Chatham Island itself are not reserved. Although the Chatham Island sow thistle has an unusual and attractive appearance, it is rarely found in cultivation. Plants are easily grown from seed but may not be long-lived.

D. Given

Comments: *Embergeria grandifolia* was originally placed in the genus *Sonchus* along with the other sow thistles but is now the sole member of the genus *Embergeria* which is endemic to the Chatham Islands. *Embergeria*: after Professor Louis Emberger (1897-1969), director of the Institute of Botany at the University of Montpellier; *grandifolia*: with large leaves.

Status: Vulnerable.

Myosotidium hortensia

kopakopa, kopukapuka; Chatham Island forget-me-not, giant forget-me-not, Chatham Island lily

Boraginaceae

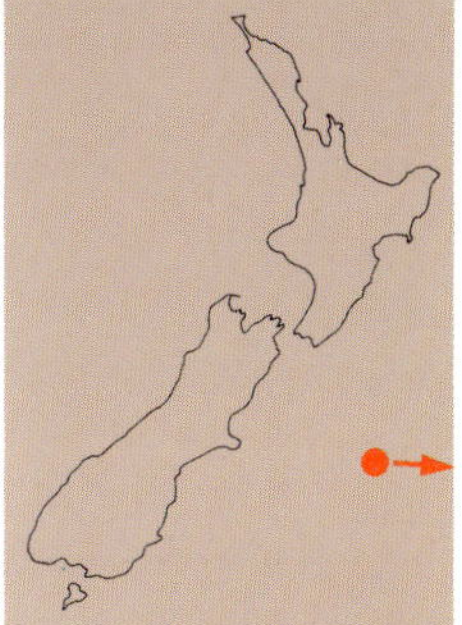

Description: Somewhat succulent herb, up to 1 m in diameter. Leaves many large, and crowded, petioles long and produced from stout rhizomes. Thick fleshy, rhubarb-like leaf laminae 13-30 cm or more long, broadly ovate to heart-shaped, margins entire, upper surfaces are deeply grooved and glossy, lower surfaces evenly scattered with minute hairs. The flowers, borne in dense pyramidal clusters, 12-15 mm in diameter, pale to dark blue, but become purplish with age. Fruit of 4 blackish, flattened nutlets, each nutlet 10-15 mm long, including the unevenly serrated wing.
Flowers: September-October; Fruit: October-April.

Distribution: *Chatham Islands. Endemic. Myosotidium hortensia* can be found on Chatham, Pitt, South East, Mangere and Little Mangere Islands, and smaller islets of the group. At one time it was described as forming an unbroken line along the seashore. This may have been an exaggeration but at least on Chatham and Pitt Islands its range has become fragmented; on Chatham Island it is now known only from a few sites.

Habitat: This coastal species has been found on sand dunes, beaches, cliff ledges, and in crannies on peat-covered rock outcrops. It seems to favour the inland edge of boulder beaches where it grows amongst kelp drift and accumulated paua shells (*Haliotis iris*).

Conservation: Although the total number of wild plants may be in the thousands, no accurate census has been made. The main factor contributing to the decline of this species has probably been grazing and trampling by introduced animals such as pigs, sheep, and cattle. Small populations occur in proposed reserves on Chatham Island, and in smaller island reserves of the Chatham group.

Myosotidium hortensia is easily grown although seed often will not germinate for up to 12 months after sowing and plants may take several years to reach flowering size; plants are sometimes attacked by cucumber mosaic virus. It is relatively widespread in cultivation and is sometimes available commercially. A white-flowered form is also becoming popular.

D. Given

Comments: The genus *Myosotidium* has only the one distinctive and outstandingly attractive species and is found nowhere else in the world. *Myosotidium*: resembling *Myosotis; hortensia*: an old name for the genus which is now called *Hydrangea.*

Status: Vulnerable.

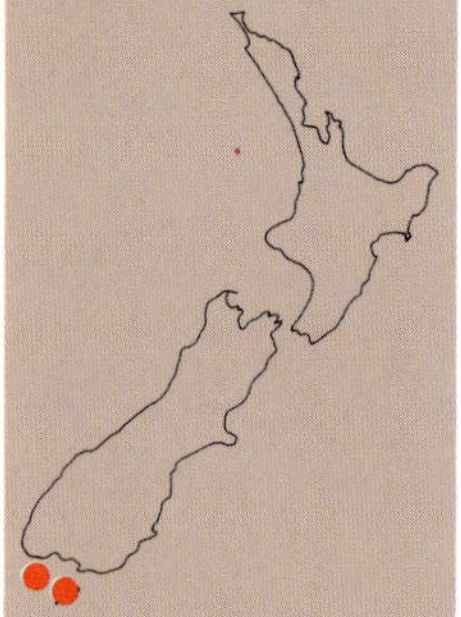

Stilbocarpa lyallii

punui

Araliaceae

Description: Robust, fleshy herb up to 1 m tall, and several metres in diameter spreading by stout, arching or prostrate stolons. Leaves with stout, hollow petioles which arise from ground level, up to 1 m long and 1 cm in diameter, laminae glossy green, roundly kidney-shaped, up to 30 cm or more in diameter, margins shallowly lobed and have triangular teeth bearing soft spines. Small, purple or occasionally yellow flowers are grouped in numerous spherical clusters. Fruit black, fleshy, shiny, 3-4 mm in diameter with 1 seed in each locule.
Flowers: November-February; Fruit: February-May of the following year.

Distribution: *Solander Islands, Stewart Island. Endemic. Stilbocarpa lyallii* is found on the Solander Islands, some islets and stacks around Stewart Island and in Foveaux Strait, and at a very few localities on Stewart Island. The species was at one time more widespread on Stewart Island and was also reported from islands in Preservation Inlet and Dusky Sound in Fiordland.

Habitat: Punui grows beneath and along the margins of coastal scrub and low forest on fertile peaty soils.

Conservation: Although this species is reasonably secure on many of the islets where it occurs, its long-term security chiefly depends on the existence of sites free of introduced browsing mammals. Rats have been known to damage punui when their population densities are high. The Solander Islands are included in Fiordland National Park. *Stilbocarpa lyallii* is occasionally seen in cultivation. It can be grown fairly readily and is probably not difficult to propagate from fresh seed.

D. Given

Comments: *Stilbocarpa lyallii* has, in the past, been placed in the genus *Kirkophytum*. Mutton-birders once stored salted mutton-bird carcasses in casks lined with the leaves of this species, or in sacks which were made from kelp and inflated with the aid of a blowpipe provided by the hollow petioles. *Stilbocarpa:* shining fruit; *lyallii*: after David Lyall (1817-1895), a member of the British Antarctic Expedition of 1839-1843. He later collected plants in the coastal areas of Fiordland and Stewart Island.

Status: Vulnerable.

Stilbocarpa robusta

Snares punui

Araliaceae

Description: Robust, fleshy herb 1 m tall, with long, prostrate stolons and short, erect stems bearing rosettes of 2-10 leaves. Leaves with solid, grooved petioles, usually up to 20 cm long and 1 cm in diameter, but sometimes up to 45 cm long, laminae kidney-shaped, up to 60 cm in diameter, and bright green and smooth on the upper surface but paler and sometimes bristly on the lower surface, margins are shallowly lobed and have triangular teeth tipped with short soft spines. Tight clusters of small pale yellow to cream flowers grouped in clusters, borne on a stout stem up to 50 cm long. Fruit black, fleshy, shiny, 3-4 mm in diameter, with 1 seed in each locule.
Flowers: November-February; Fruit: December-February of the following year.

Distribution: *Snares Islands. Endemic. Stilbocarpa robusta* grows on the Snares. It has also been reported from Little Solander Island but this record has not been confirmed recently.

Habitat: On the Snares Islands *Stilbocarpa robusta* is found in open forest, clearings, *Hebe elliptica* scrub, and grassland. It seems to prefer shady sites or areas where the soil is moist.

Conservation: Thousands of plants grow on the Snares Islands which are protected as a Nature Reserve. Although this species seems reasonably secure at present, its small geographic range and palatability to introduced grazing mammals makes *Stilbocarpa robusta* very vulnerable. If this species is to survive in the wild it is probably essential that grazing mammals and rats do not get on to the Snares Islands. There are no confirmed records of *S. robusta* occurring in cultivation.

Comments: This species, together with *Stilbocarpa lyallii*, was once placed in the genus *Kirkophytum*. *Stilbocarpa*: shining fruit; *robusta*: strong, well built.

Status: Vulnerable.

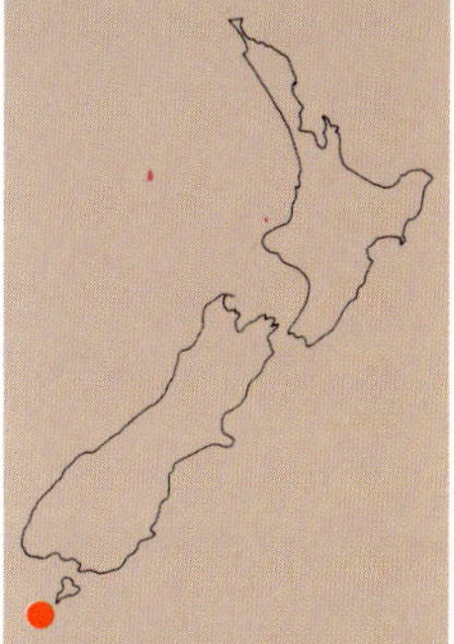

R. de Hamel

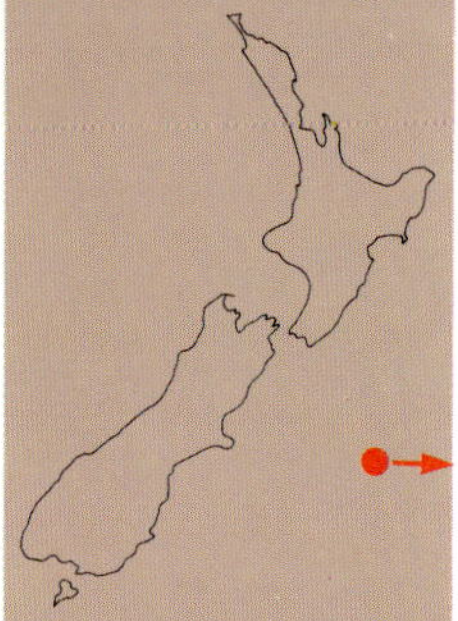

Aciphylla dieffenbachii

Dieffenbach's speargrass, Dieffenbach's spaniard

Apiaceae

Description: Tussock-like herb with many leaves up to 60 cm, arising from the base of the plant. Leaves 8-30 cm long, of 4-5 pairs of pinnae each divided into somewhat flaccid linear segments with short spines at tip. Stout flower stems up to 1 m tall, 4 cm in diameter. Flowers small, in numerous round clusters, cream; male and female flowers on separate plants. Fruit dry, ellipsoid and golden yellow, (schizocarp), 10-15 x 6-10 mm, each splitting into 2 single-seeded mericarps with 2 or 3 wings. Flowering and fruiting periods not known.

Distribution: *Chatham Islands. Endemic.* This speargrass occurs on Chatham Island, Mangere Island, Pitt Island, and Little Mangere Island in the Chatham group. Although the species is endemic to the Chatham Islands, a small population has been deliberately naturalised on an islet off Stewart Island.

Habitat: *Aciphylla dieffenbachii* always grows close to the sea and is usually found on coastal cliffs and steep colluvial slopes, frequently overlying rocks of volcanic origin.

Conservation: Hundreds of plants exist on Mangere and Pitt Island but the generally small size of the colonies found on Chatham Island gives cause for concern. Two reserves have been established on Chatham Island. Most of the plants found growing outside these reserves are inaccessible to stock. This attractive species is occasionally seen in public and private gardens. Seed treated in the laboratory germinated after 6 months at 5°C.

Comments: This species is sometimes regarded as the sole member of the genus *Coxella*. *Aciphylla*: sharp leaf; *dieffenbachii*: after Dr Ernst Dieffenbach (1811-1855), a German doctor and naturalist who stayed in New Zealand from 1839 to 1841.

Status: Vulnerable.

G. Walls

Aciphylla traversii

Chatham Island speargrass, Chatham Island spaniard

Apiaceae

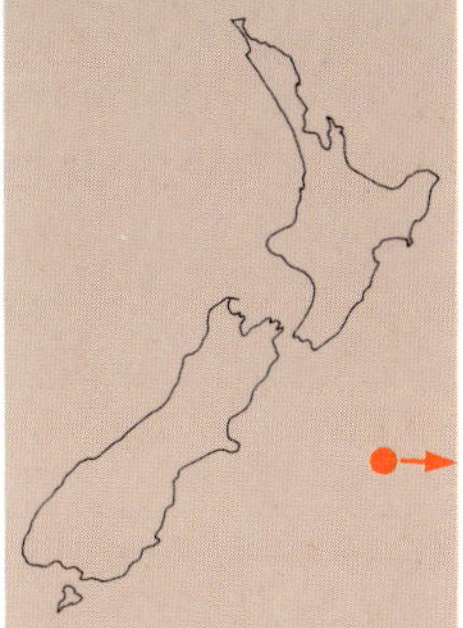

Description: Stout, tussock-like herb up to 1 m tall with many leaves arising from the base of the plant. Leaves 15-100 cm long, of several pairs of linear pinnae which are each 10-40 cm x 4-12 mm, and terminate in a stiff, sharp spine. Young plants have stiff, erect leaves, becoming more flaccid as the plant matures. Flower stems up to 1 m tall. Flowers small, in numerous rounded or open clusters, cream; male and female flowers on separate plants. Fruit of oblong schizocarps; 8-10 x 2-3 mm, each and splitting into 2 single-seeded mericarps with 2-4 wings.
Flowers: November-April; Fruit: March-May.

Distribution: *Chatham Islands. Endemic.* Chatham Island speargrass occurs only on Chatham Island where it is virtually confined to the Southern Tablelands.

Habitat: Small, scattered colonies or isolated individuals grow on peaty-sandy soils and in bogs dominated by *Sporadanthus traversii* and *Olearia semidentata*. It seems to prefer open, gently sloping sites and peaty lake margins, although it has also been found on forest margins, below canopy gaps, and in scrubland on peaty soils.

Conservation: Although it may never have been particularly abundant, numbers of Chatham Island speargrass have been reduced by a combination of burning and cattle browsing. The largest plants are on infrequently burnt sites, but young plants are sometimes common where there has been recent burning. The species is known to be in two reserves and a small, private, fenced enclosure in the north-eastern corner of the island. This private "reserve" was established in the 1940s from seed obtained on the Southern Tablelands.
A few plants of this species are currently grown in botanic and experimental gardens. Up to 2 years or more may elapse before germination occurs.

Comments: *Aciphylla*: sharp leaf; *traversii*: after Henry Travers (1844-1928), who made some of the earliest botanical collections on the Chatham Islands. He was the son of William Travers, a District Court Judge, M.P., and naturalist.

Status: Vulnerable.

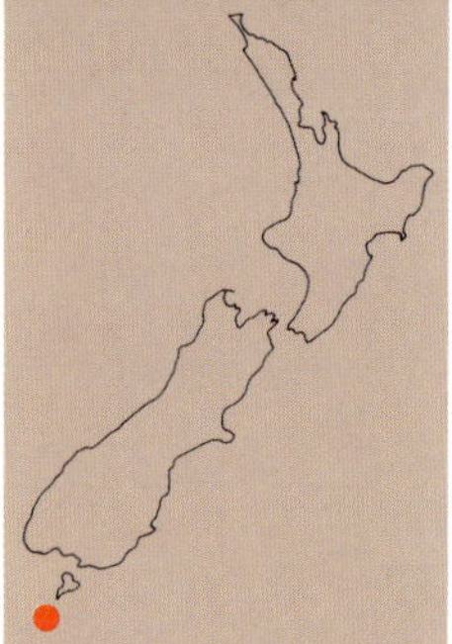

Anisotome acutifolia

Apiaceae

Description: Very stout herb up to 2 m high with short, creeping stolons up to 30 cm long. Leaves tough, arranged in a rosette, petioles up to 30 cm, laminae 15-40 x 14-30 cm, ovate to triangular; each lamina divided into 6-8 pairs of primary pinnae up to 15 x 8 cm wide, these again divided into secondary pinnae 3-5 x 1-2 cm, and tertiary pinnae up to 2 cm long, tertiary pinnae deeply or shallowly cut into irregular segments. Flower stem up to 1 m high, deeply furrowed. Flowers in rounded clusters, 4-5 mm in diameter, petals white with pale purplish pink margins and an occasional purplish pink central stripe. Fruit 4-5 mm long, splitting into 2 deeply ribbed mericarps with distinct wings.
Flowers: December-January; Fruit: January-February.

Distribition: *Snares Islands. Endemic. Anisotome acutifolia* is found on Northeast and Broughton Islands.

Habitat: This herb usually grows in coastal grassland near forest margins.

Conservation: The Snares Islands, which are now a Nature Reserve, support about 1000-3000 plants. There are two large patches on Northeast Island and several minor ones on Northeast and Broughton Islands. Although tens of thousands of seeds are produced each year, fewer than 20 seedlings were found on Northeast Island in the summer of 1984-85, and reproduction is predominantly vegetative. The flower and seed heads are often heavily infested with stem-mining moth larvae and aphids but it is unlikely that insect pests alone are responsible for the small number of seedlings. Very few specimens of *A. acutifolia* are known to be in cultivation. The species should be easy to grow from seed.

Comments: *Anisotome*: unequally cut, referring to the unequal carpels; *acutifolia*: with pointed leaves.

Status: Vulnerable.

D. Horning

Astelia chathamica

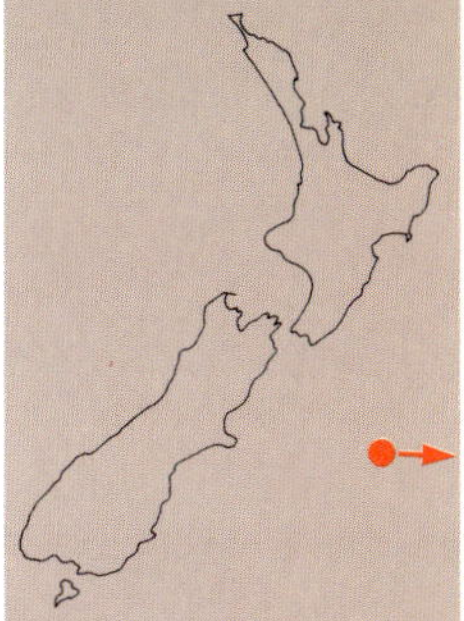

Chatham Island kakaha

Liliaceae

Description: Tussock-like clumps of many overlapping sword-shaped leaves which arise at ground level, up to 60-200 x 4-10 cm. Upper surface of leaves silvery green with a metallic sheen, covered by a thin, clear, skin-like layer which lifts off the old leaves in strips, lower surfaces of the leaves clad in a pale grey satiny layer of small, flat scales. Flower stems short, 2- or 3-branched usually concealed by the leaves. Flowers numerous, pale greenish-white, about 8 mm long, partially hidden by bracts. Subglobose, fleshy berries are orange, about 10 mm in diameter.
Flowers: October; Fruit: March-June.

Distribution: *Chatham Islands. Endemic*. Once abundant on Chatham Island, this species is now known mostly from small colonies widely scattered throughout Pitt and Chatham Islands.

Habitat: Chatham Island kakaha once grew in a variety of moist sites including forest floors, swamps, stream margins, and rocky areas, but is now found mainly on rocky sites in forest and scrub, on cliffs and steep slopes and along lake margins and streams where it is out of the reach of livestock.

Conservation: This plant usually occurs in small populations of about 20 or 30 plants. Its restricted distribution is probably a result of burning and grazing pressures which deplete this very palatable species. Present reserves on the Chatham Islands may not adequately protect this species in the long-term. *Astelia chathamica* is readily grown from fresh seed or offsets and is occasionally grown in public and private gardens in New Zealand. It is sometimes available commercially.

Comments: *Astelia*: lacking a stem; *chathamica*: of the Chatham Islands.

Status: Vulnerable.

D. Given

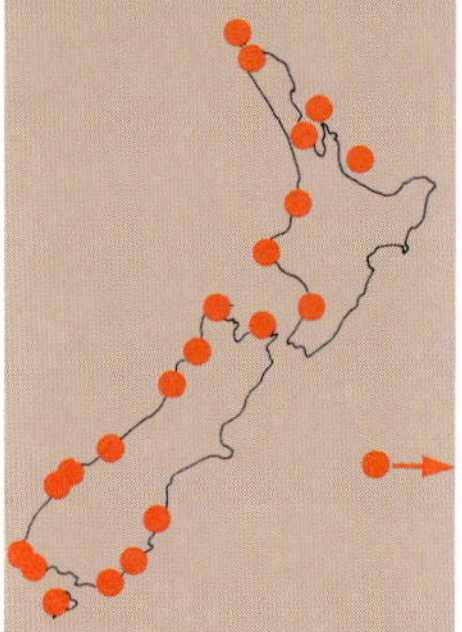

Euphorbia glauca

waiu-atua, wai-o-Kahukurai; New Zealand sea spurge, Maori spurge, sand milkweed

Euphorbiaceae

Description: Soft herb up to 1 m, stems erect, rhizomes creeping, sap milky. Leaves bluish green, oblanceolate to obovate or somewhat elliptic, margins entire, 2-12 x 1.5-2.5 cm. Juvenile leaves, and often the stems, have a reddish tinge. Flowers minute in small inflorescences produced near the tips of the stems, each surrounded by a red cup-shaped structure bearing purple, crescent-shaped glands around its rim. Fruit a pendulous, 3-lobed capsule which changes from green to pale brown as it ripens, seeds smooth, pale yellow to purplish brown, 3-7 x 3 mm.
Flowers: October-February; Fruit: December-April.

Distribution: *North Island, South Island, Stewart Island, Chatham Islands. Endemic. Euphorbia glauca* is sporadically distributed throughout the North and South Islands of New Zealand, probably reaching its greatest abundance in Fiordland. It is also known from a few sites on Stewart Island and two sites on Chatham Island.

Habitat: Although open sand dunes are the most usual habitat of this coastal species, it has also been found on gravel, in rocky places, on coastal banks, and in seepages near the shore. In some areas it shows a preference for limestone talus.

Conservation: Numerous colonies of *Euphorbia glauca* are known, but most are small and often do not extend over more than a few square metres. Induced sand instability, trampling by animals, and overtopping by scrub weeds such as tree lupin have possibly led to the disappearance of this species from many dune systems. Coastal forestry and farming may have affected its habitat in some parts of New Zealand. *E. glauca* occurs in several reserves. It can be readily propagated from seed or offsets and is sometimes found in cultivation.

D. Given

Comments: *Euphorbia glauca* is the only spurge or milkweed which is native to New Zealand. *Euphorbia:* from Euphorbos, physician to the King of Mauretania in about 24 B.C.; *glauca*: dull green to greyish blue.

Status: Vulnerable.

Pratia physaloides

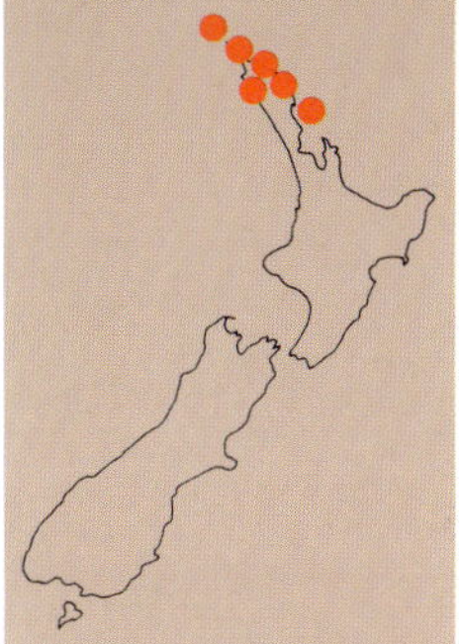

koru, oru

Lobeliaceae

Description: Erect to sprawling, leafy, shrub-like herb, up to 1 m tall and resembles a lax *Hydrangea bush*. Leaves confined mainly to the tips of the branchlets, petioles up to 10 cm, laminae 5-20 x 3-10 cm, ovate to ovate-elliptic or occasionally elliptic with finely to coarsely toothed margins. Flowers pale to deep violet-blue or sometimes whitish, 3-5 cm long, produced near the tips of the stems. Fruit of subglobose, bright blue to whitish berries, 10-15 mm in diameter, containing numerous small, orange-brown seeds. Flowers and fruit can be found throughout most of the year.

Distribution: *North Island, Three Kings Islands. Endemic*. Koru occurs on the Three Kings Islands and several other northern offshore islands, and is locally distributed throughout the northern part of the North Auckland region.

Habitat: Plants of this species usually grow in coastal and lowland forest and are most often found along the banks of streams, on damp banks, and along the forest margins. It is locally abundant in forest growing on talus slopes on some offshore islands.

Conservation: Koru has been recorded from numerous localities but large North Auckland populations are now rare due to the opening up of forest. Goats and possums pose a severe threat to the survival of several North Auckland colonies. Although large populations occur in Te Paki Farm Park and in an ecological area and Forest Park within Warawara State Forest, the offshore islands may offer the most secure sites for the long-term survival of this species. *Pratia physaloides* is sometimes seen in cultivation in the northern half of the North Island. It is easily grown from seed and cuttings, but is susceptible to attack by slugs and snails.

Comments: Its blue flowers and berries makes this one of New Zealand's most unusual plants. The species is so distinctive some taxonomists regard it as the only member of the segregate genus *Colensoa*. *Pratia:* after C.L. Prat Bernon (?-1817), a French botanist; *physaloides*: resembling *Physalis* (Cape gooseberry).

Status: Vulnerable.

W. Rennie

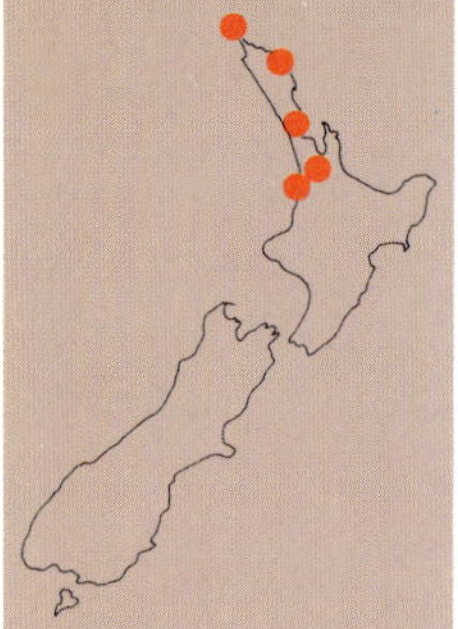

Ranunculus urvilleanus

Ranunculaceae

Description: Slender, tufted herb up to 80 cm or more, stems and leaves sparsely covered in soft, white hairs. Lower leaves with slender petioles up to 35 cm, laminae 3-5 or sometimes up to 10 cm long, divided into 3 stalked leaflets, each broadly ovate leaflet deeply-lobed, margins coarsely toothed, 3-5 cm long. Upper stem leaves similar but smaller. Erect, branching flower stems with several bright yellow flowers, each 8-15 mm in diameter. Fruiting head globose, of 30-50 brown, compressed achenes, 3 x 2 mm, including the hooked beak.
Flowers: October-January; Fruit: November-February.

Distribution: *North Island. Endemic. Ranunculus urvilleanus* has been collected throughout the northern region of the North Island - mainly in the eastern district between North Cape and Morrinsville - and also from Little Barrier Island. It is known to survive in few locations at present - four between Te Paki and Russell, on Tiritiri Matangi Island, and at several sites in South Auckland (where further sites are likely to exist).

Habitat: *Ranunculus urvilleanus* has been found in damp sites in coastal and lowland areas, along swamp and stream margins, and on wet ground amongst scrub.

Conservation: About 80 plants have recently been found at 3 sites in 1 locality and at least 1 other population consists of numerous plants. Little is known about the other three locations which have not been inspected for several years. The species may benefit from a low level of grazing by livestock, since the removal of sheep and cattle from some areas appears to have reduced the number of bare open areas required for the establishment of young plants.
Ranunculus urvilleanus is not known to be in cultivation.

Comments: *Ranunculus*: from the Latin name for a small frog or tadpole, alluding to the swampy habitats preferred by some members of this genus; *urvilleanus*: after Jules Dumont D'Urville (1790-1842), a French naval officer and botanist.

Status: Vulnerable.

E. Cameron

Celmisia adamsii

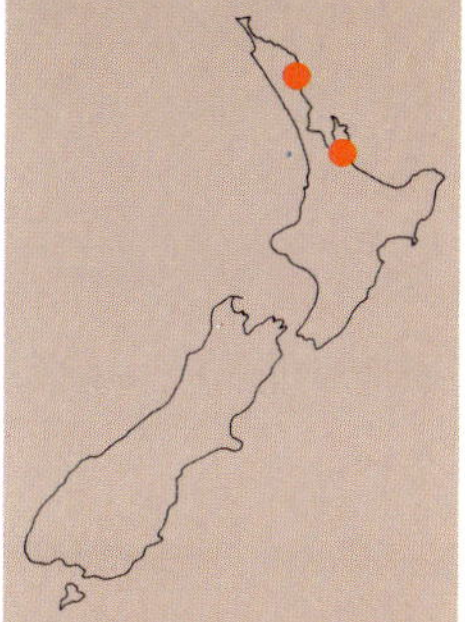

Asteraceae

Description: Small to medium sized tufted herb, with rosettes of silvery green leaves which are clad in soft white hairs on their lower surface and may be up to 38 x 3 cm wide. Each leaf is narrowly oblong-lanceolate to almost linear with small, widely spaced teeth along its margins, and often has a short point at the tip. The slender, hairy, and sometimes flexuous flower stems are usually up to 30 cm tall. They bear flower heads which are 3-5 cm in diameter and consist of yellow disc-florets surrounded by white ray-florets. The achenes are about 4 mm long and are crowned by slender white hair-like bristles up to 5 mm long.

Variety *rugosula* is very similar to var. *adamsii*, but stouter and shorter overall and its wrinkled leaves are more silvery, flower stems are covered in more dense hairs and flower heads are larger (about 5 cm in diameter).

Flowers: December-February; Fruit: March-April.

Distribution: *North Island. Endemic.* Var. *adamsii* has been recorded from a few sites on Coromandel Peninsula; var. *rugosula* occurs at about 10 adjacent sites near Whangarei Harbour.

Habitat: *Celmisia adamsii* is usually found on steep rock faces, rocky crags, and shaded cliffs, and shows some preference for sites where the soil is moist or semi-shaded, well drained, and only seasonally dry.

Conservation: *Celmisia adamsii* var. *adamsii* seems at one time to have been slightly more widespread on Coromandel Peninsula, but appears to have declined through disturbance by goats, trampers, and plant collectors. Plants in the wild may still be threatened by these factors even though 2 populations totalling about 250 plants are within a forest park. Var. *rugosula* was once more widespread north of Whangarei Harbour, but clearance of scrubland for pasture, animal trampling, and specimen collecting has reduced its range and abundance. Two populations, each with several hundred plants are within a scenic reserve. *C. adamsii* is occasionally found in cultivation. Seed usually germinates readily and mature plants can be divided successfully, but the species may not be very long-lived.

Comments: *Celmisia*: from Celmis, one of the high priests of Cybele; *adamsii*: after James Adams (1839-1906), a well known New Zealand botanist who was headmaster of the Church of England Grammar School in Parnell and Thames High School; *rugosula*: wrinkled.

Status: Vulnerable (both varieties).

E. Cameron

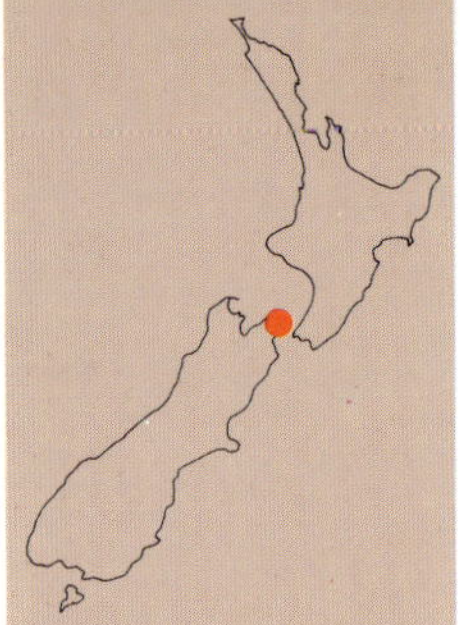

Celmisia macmahonii var. macmahonii

Asteraceae

Description: Woody-based herb forming cushions up to 60 cm in diameter. Leaves narrowly oblanceolate to narrowly oblong, arranged in dense rosettes, 20-35 x 6-8 mm, short point at the tip, both upper and lower surfaces covered with soft, greenish white hairs which turn chocolate-buff when dried. Flower stems 8-15 cm long and densely covered in similar silky hairs. Flower stem terminates in a flower head (capitulum), 20-40 mm in diameter and consists of yellow disc-florets surrounded by white ray-florets. Fruit of grooved achenes, about 4 mm long and crowned by slender white hair-like bristles about 6 mm long.
Flowers: December-January; Fruit: February-April.

Distribution: *South Island. Endemic.* The only known population of *Celmisia macmahonii* var. *macmahonii* grows on a peak in the Marlborough Sounds.

Habitat: This alpine plant is confined to ledges and crevices on bluffs and rock outcrops near the treeline.

Conservation: Feral goats have devastated the area where this plant occurs since the early 1900s, and continue to threaten the species' survival. Fortunately the entire known population is within a scenic reserve. Several attempts to grow this species have been made, but it has proved difficult to propagate and maintain in cultivation.

Comments: The only other form of this species is the more abundant var. *hadfieldii* which occurs further south on the Richmond Range. It forms a very tight cushion and has narrower leaves 25-30 x 3-5 mm wide.
Celmisia: from Celmis, one of the high priests of Cybele; *macmahonii*: after J.H. Macmahon who first collected this species.

Status: Endangered.

D. Given

Gunnera hamiltonii

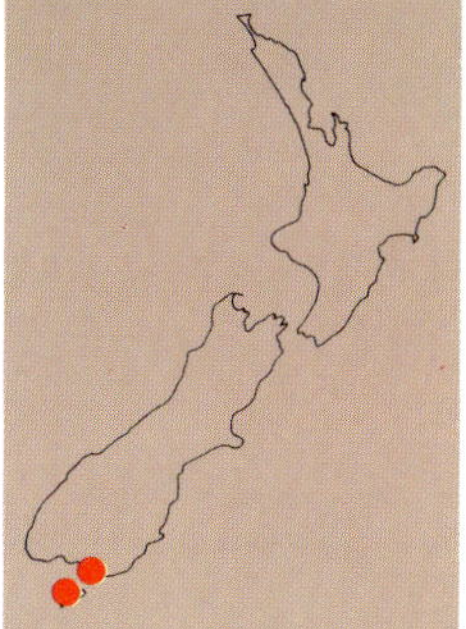

Gunneraceae

Description: Ground-hugging herb forming patches of overlapping rosettes or growing as solitary rosettes reaching 10 cm in diameter. Leaves of tough, triangular laminae, on wide, reddish petioles, brownish green to greyish green tinged with red, margins regularly toothed, each 1-5 x 0.5-2.5 cm. Flowers numerous, tiny, on erect stems up to 12 cm tall; male and female flowers on different plants. Unfertilised drupes are red, fleshy, and about 3 mm long, but fresh fertilised drupes have yet to be examined.
Flowers: August-October; Fruiting period not known.

Distribution: *Stewart Island. Endemic.* Only five colonies of *Gunnera hamiltonii* have been found. One of these, which grew beside the Oreti River near Invercargill, on the southern coast of the South Island, was destroyed in the late 1960s by the combined effects of grazing, weeds such as lupins, and the removal of sand. Fortunately, a new population at another site near Invercargill was discovered early in 1987. Three other colonies grow in two localities on Stewart Island.

Habitat: *Gunnera hamiltonii* is found in sand-dune hollows near the sea. It grows in damp sandy sites in the narrow transition zone between unstable dunes and stabilised scrub-covered sand.

Conservation: Plants at the site near Invercargill grow in an area which is grazed by cattle. The largest patch covers 5 x 6 m. The influx of competing weeds is a continuing but manageable threat to the three Stewart Island colonies which together occur over only a few hundred square metres. The material in these populations may be all male but some is undetermined as to sex. Plants from both Stewart Island and Invercargill are in cultivation and there is a chance of obtaining good seed by fertilising female Invercargill plants using Stewart Island males. Some plants at one Stewart Island site have been transplanted to another secure location nearby and the area is reserved. An attractive rock garden plant, *Gunnera hamiltonii* is grown in both public and private gardens and is occasionally available commercially. It establishes readily from single rosettes.

D. Given

Comments: Two wild populations of *Gunnera hamiltonii* grow amongst *G. prorepens,* a vigorous relative of similar growth form and appearance. *G. hamiltonii* is readily distinguished by its wide, reddish petioles and triangular laminae; *G. prorepens* has more ovate laminae. *Gunnera*: after Johan Gunnerus (1718-1773), a Norwegian bishop and botanist; *hamiltonii*: after W. S. Hamilton of Southland.

Status: Endangered.

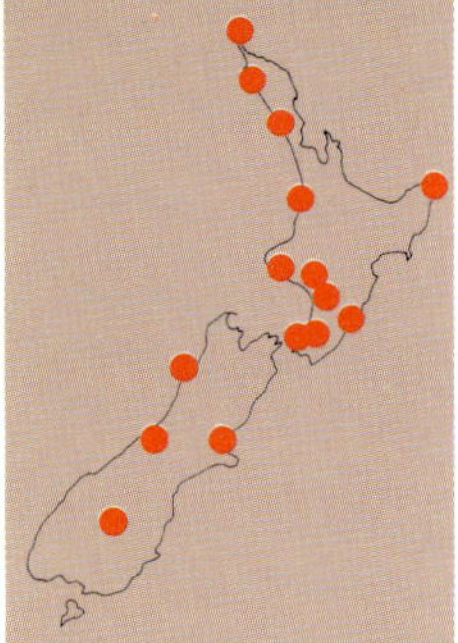

Mazus pumilio

dwarf mazus, dwarf false musk

Scrophulariaceae

Description: Small herb. Leaves arranged in erect tufts or close rosettes, laminae 1-7.5 cm long, obovate-spathulate in shape, margins entire or slightly toothed, surfaces may sometimes be sparsely hairy. Flower stems slender, each bearing 1-6 flowers, 6-12 mm long, and white or sometimes light blue with yellow throats. Seed capsules compressed and globose, encased in persistent calyx, each containing numerous minute seeds.
Flowers: November-February; Fruit: December-March.

Distribution: *North Island, South Island. Also found in Australia and Tasmania.* Dwarf mazus is scattered from North Cape to Lowburn in Otago.

Habitat: Dwarf mazus has been recorded from swamp and stream margins, boggy ground, river flats beneath tawa and kahikatea, rocky coastal headlands amongst coastal pasture grasses, and sandy turf on the inland margin of a beach.

Conservation: Very little is known about the occurrence of dwarf mazus in reserves throughout New Zealand and no reliable estimate of the total number of plants is available. Competition with exotic plants is probably the major threat to this species. It is easily propagated by division of single plants, but very few plants are in cultivation.

Comments: *Mazus radicans*, which occurs in the North Island and may be found in similar habitats, is distinguished from *M. pumilio* by its stouter habit, longer flowers which are 12-20 mm long, and shorter leaves which are 2-5 cm long. *Mazus*: teat, referring to two small teat-like protuberances in the throat of the flower; *pumilio*: a dwarf.

Status: Vulnerable.

R. Lamberts

Myosotis albosericea

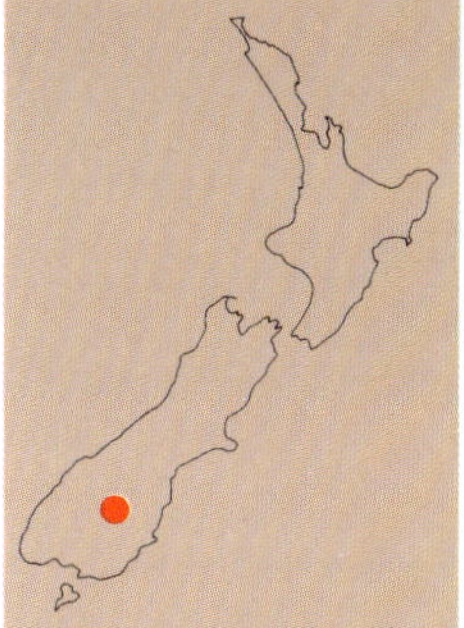

Boraginaceae

Description: Low, branching herb forming small cushions. Leaves silvery or greyish, narrow spathulate, up to 3 cm long, arranged in rosettes, both upper and lower surfaces covered in a dense layer of stiff, straight hairs pressed closely against the leaf. Flowers lemon-yellow and about 5 mm in diameter, anther tips barely protruding above the small scales found at the entrance to the inner tubular portion of the flower. Pointed nutlets ovoid, about 1.7 mm long.
Flowers: December-February; Fruit: February-April.

Distribution: *South Island. Endemic. Myosotis albosericea* has been collected only from the Dunstan Range and Cromwell Gorge in Central Otago. Most collections were made last century and early this century, and by the early 1900s the species was thought to be almost extinct. In 1976 it was rediscovered at a single site in the same region.

Habitat: All the plants known in the wild grow on exposed, eroded pavements of shattered schist rocks at about 1500 m above sea level.

Conservation: Only one population of several hundred plants has been found in the wild, and it appears that this species has always have been very local in distribution. The site is partly under pastoral lease, but the effects of grazing on the species are not known. A vehicle track adjacent to the population constitutes some threat to the survival of *Myosotis*. Any maintenance of this track will require careful planning and supervision to prevent the excavation or burial of adajcent plants. A few specimens are in cultivation.

Comments: The genus *Myosotis* is interesting because a high percentage of New Zealand species are of very local distribution. *Myosotis*: mouse ear, referring to the hairy leaves; *albosericea*: with white hairs.

Status: Endangered.

D. Given

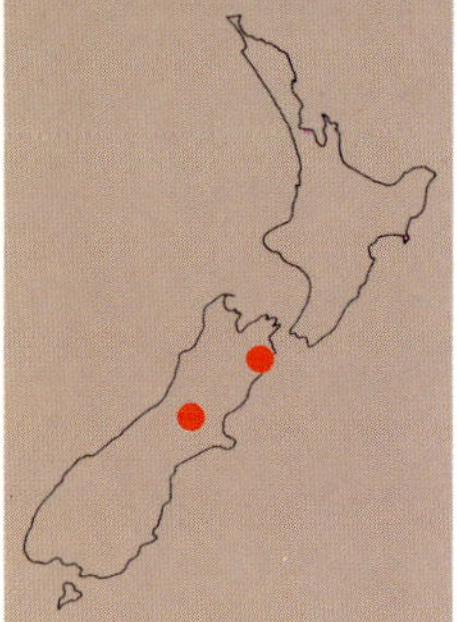

Myosotis colensoi

Castle Hill forget-me-not

Boraginaceae

Description: Low herb with numerous leaves forming small cushions or a loose mat of rosettes. Leaves spathulate to oblanceolate, up to 4 cm long, upper surface of each leaf has straight hairs of unequal length pressed closely against the leaf whereas the hairs on the lower surface are more equal in length, and less numerous or sometimes absent. Flowers white, up to 12 mm in diameter (larger than those of most other New Zealand forget-me-nots), solitary or grouped in small clusters. Anther tips protrude above the small scales found at the entrance to the inner tubular portion of the flower. Fruit of shiny black nutlets, oblong, about 1.2 mm long.
Flowers: October-December; Fruit: December-January.

Distribution: *South Island. Endemic. Myosotis colensoi* grows at several sites in the Castle Hill basin in Canterbury, and has also been recently discovered on the Chalk Range in Marlborough.

Habitat: On limestone talus, most common in open areas of relatively fine debris. At Castle Hill it is often associated with *Wahlenbergia brockiei* and *Poa acicularifolia*.

Conservation: The greatest threat to the survival of the several hundred plants growing in the Castle Hill area may be from aggressive weeds such as oxeye daisy and *Hieracium* spp., and a reduction in areas of open habitat suitable for the species. A large population is included in a nature reserve. *Myosotis colensoi* can be readily grown from seed and is in limited cultivation in private and public gardens. Its relatively large flowers make this an attractive rock garden plant.

D. Given

Comments: *Myosotis colensoi* has also been reported from Pourakino Valley in Southland, but the locality itself has not been checked.
Myosotis: "mouse ear", referring to the hairy leaves; *colensoi*: after William Colenso (1811-1899), a missionary and a keen botanist who collected plants throughout the North Island.

Status: Vulnerable.

Myosotis traversii var. cinerascens

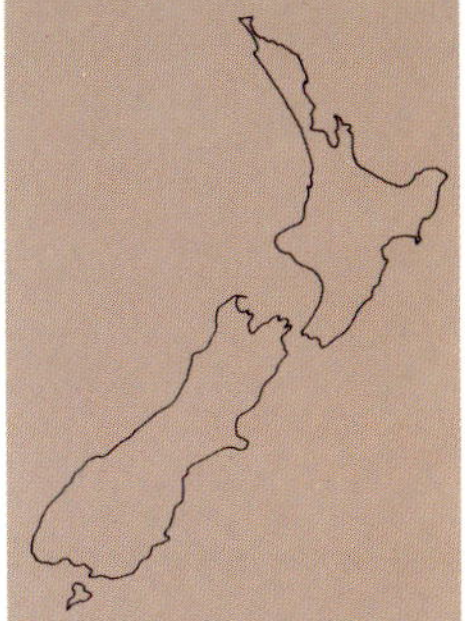

Boraginaceae

Description: Small herb forming 1 to several loose rosettes of spathulate leaves, up to 3 cm long. Upper surfaces of these rosette leaves are covered in short, overlapping hairs of unequal length whereas the hairs on the lower surfaces are all short. The lower surfaces of the uppermost leaves on the stems have similar short hairs interspersed with longer hooked hairs. Flowers lemon-yellow to white, about 4 mm in diameter, anther tips protrude above the small scales found at the entrance to the inner tubular portion of the flower. Fruit of thin, shiny, oblong, dark brown nutlets; 2-2.5 mm long.
Flowering and fruiting periods not known.

Distribution: *No wild populations known. Endemic.* This species is known only from a few specimens collected early this century in the Castle Hill basin, in the South Island. Recent searches have failed to relocate it.

Habitat: Brief notes on herbarium sheets suggest that *Myosotis traversii* var. *cinerascens* was found only on limestone debris.

Comments: The arrangement of the leaf hairs distinguishes this variety from other, more widespread, forms of the species. *Myosotis*: mouse ear, referring to the hairy leaves; *traversii*: after William Travers (1819-1903), a District Court Judge, M.P., and naturalist; *cinerascens*: greyish.

Status: Presumed extinct.

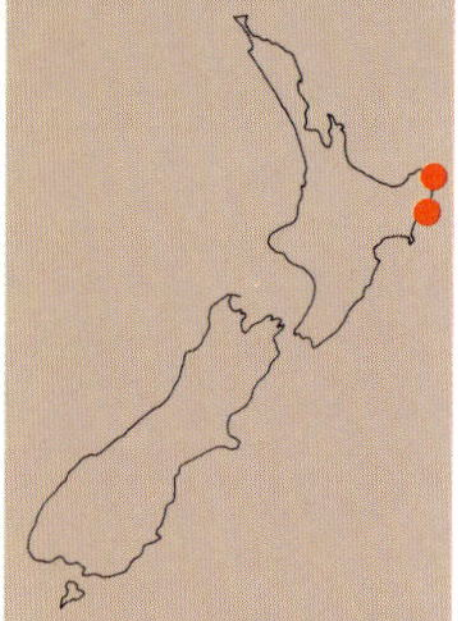

Plantago spathulata subsp. picta

Plantaginaceae

Description: Low herb with leaves arranged in a flat rosette up to 17.5 cm in diameter. Leaves oblanceolate to spathulate, 2-8 x 1-2.5 cm wide, sometimes a few widely spaced teeth along margins, small groups of sparse hairs occur near the margins on the upper surface of the leaves. Flower stems are covered in brown hairs, up to 12 cm long. Flowers minute, in dense spikes 1-3.5 cm long, each flower is associated with a bract 2.5-5 mm long, which is dark in the centre, with pale, membranous margins and fringed with short hairs. Fruit of ellipsoid capsules up to 3 mm, usually containing 4 flat, ellipsoid seeds 1.5-3 mm long. Flowers: November-December; Fruit: December-January.

Distribution: *North Island. Endemic.* Six populations of *Plantago spathulata* subsp. *picta* are known from sites between Gisborne and Cape Runaway.

Habitat: Most populations of this coastal species are found on rapidly eroding cliffs of soft mudstone (papa), but at one site it grows in seepages and small cracks on a steep coastal slope of volcanic rock.

Conservation: Several hundred plants have been discovered at one site, but many of these plants were seedlings which may not all survive to maturity. A total of 200-300 plants survive at the remaining localities. The species usually grows in a very unstable habitat and it is possible that it may only occupy particular sites for short periods. The main long-term threats to the survival of this *Plantago* in the wild include the unstable nature of the plant's habitat, browsing by goats and stock, and competition with weeds. One population has been depleted within the last 25 years through the erosion of the soft mudstone cliffs on which it grows. Only about 100 plants from a single site are protected within a formal reserve. Although this species is not difficult to grow, only a few plants are cultivated in private and botanic gardens.

R. Lamberts

Comments: Rather similar plants have been reported from coastal sites near Westport and in north-west Nelson in the South Island, but their relation to authentic subsp. *picta* is not known. *Plantago*: sole of the foot, referring to the broad leaves which are flattened against the ground; *spathulata:* spoon-shaped, referring to the leaves; *picta*: painted or variegated.

Status: Endangered.

Acaena pallida

sand piripiri; sand bidibid

Rosaceae

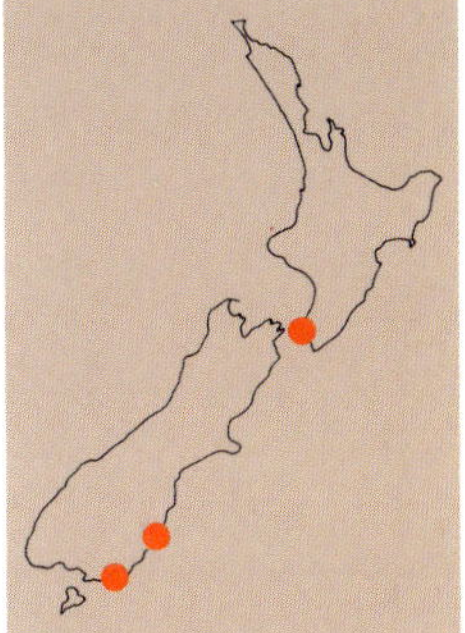

Description: Low, stout herb, stems prostrate, up to 1 m or more, clothed in soft, shaggy hairs, but seldom seen as usually covered by loose sand. Branches clad in similar hairs and up to 15 cm or more. Leaves up to 10 cm or more, divided into 9-13 leaflets with glossy green upper surfaces, very pale lower surfaces, and coarsely toothed margins. Leaflets near the tip of each leaf, obovate-oblong to oblong, 1-3 x 0.5-1.5 cm. Flower stems stout, leafless, up to 15 cm or more, clad in soft shaggy hairs. Each flower stem supports a globose flower head bearing pale brown spines which are sometimes tinged with red. Flower heads about 4 cm in diameter, including the spines.
Flowers: October-December; Fruit: December-April.

Distribution: *North Island, South Island. Also found in south-east Australia*. The species occurs at several sites along the coastline near Wellington in the North Island, and near Dunedin and the mouth of the Catlins River in the South Island. It has also been recorded from Bluff in Southland and from Stewart Island, but these records have not been confirmed recently.

Habitat: *Acaena pallida* is usually found towards the inland margin of beaches where it inhabits coastal dunes of loosely packed sand and slopes of fine gravel. It may form mats covering up to 1 m^2 or straggle over neighbouring plants.

Conservation: Remaining populations are threated by disturbance of their habitat and invasion by introduced plants which compete for open sites with loosely packed sand. Only one of the sites where it occurs is reserved and only a very few specimens are in cultivation. The seed germinates easily and growth is rapid but, although the species is perennial, it may not be long-lived.

Comments: *Acaena*: a thorn; *pallida*: pale, referring to the leaflets. However, this name may be misleading. Fresh material has dark green leaflets.

Status: Vulnerable.

P. Johnson

Lepidium banksii

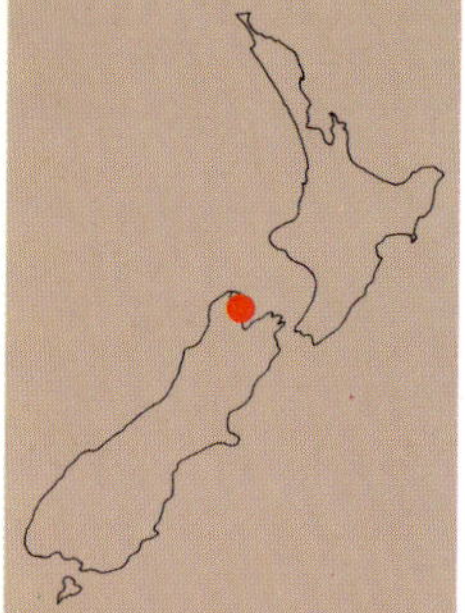

Brassicaceae

Description: Stout herb similar to *Lepidium oleraceum* (p. 82) but shorter and with a less spreading habit. Leaves usually 1.5-5 x 0.5-0.8 cm, oblanceolate-spathulate, margins finely toothed, petioles broad, flat and about equal in length to the laminae. Flowers small, white, about 1.5 mm long, clustered at tips of branchlets. Fruit of a dry silicle, broadly ovate, heart-shaped at the base, notched at apex, 5-5.5 x 4-5 mm. The 2 brown, broadly ovoid seeds within each silicle are about 2.5 mm long.
Flowers: November-March; Fruit: December-April.

Distribution: *South Island. Endemic.* Since its discovery last century *Lepidium banksii* has occasionally been collected from the Nelson region. The most recent material was collected from two sites near Totaranui in 1985 and one near Nelson.

Habitat: Wild colonies inhabit rocky coastal sites just above the high tide mark and adjacent to the roosting sites of sea birds.

Conservation: Two populations, both of which are very small, are within Abel Tasman National Park. *Lepidium banksii* can be easily grown from seed. A few plants are in cultivation.

Comments: *Lepidium*: a scale, referring to the shape of the silicle; *banksii*: after Sir Joseph Banks (1743-1820), who accompanied Lt James Cook as a scientist aboard the *Endeavour* and funded much of the scientific work of the expedition.

Status: Endangered.

P. Brooke

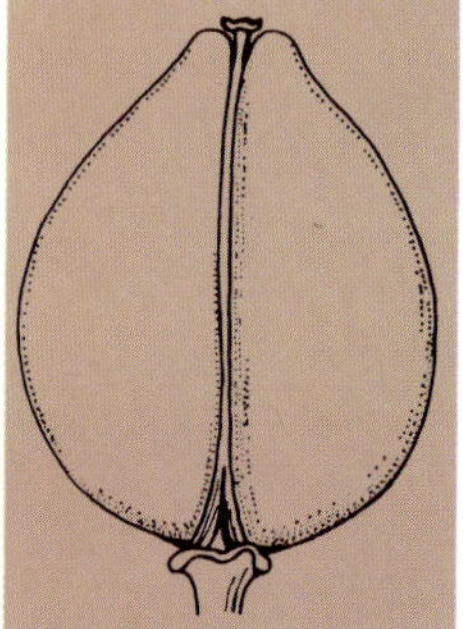

seed

D. Given

Lepidium kirkii

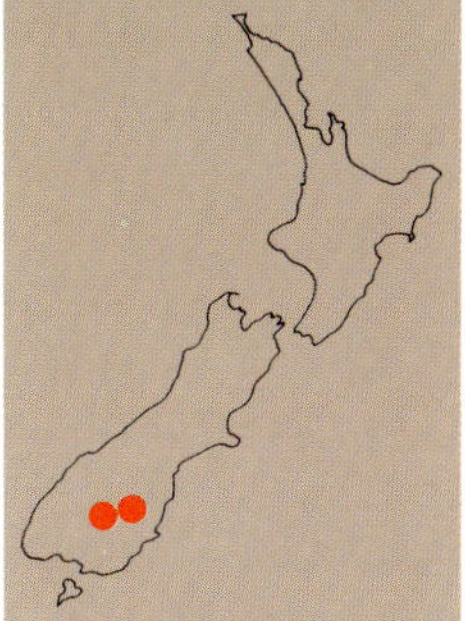

Brassicaceae

Description: Slender herb with thread-like, trailing stems reaching up to 10 cm. Its very narrow rosette leaves are linear-spathulate, margins entire, up to 3 cm x 6 mm. Minute white flowers in short clusters which elongate up to 3 cm as the fruit develops. Fruit of flattened, ovoid silicles, very shallowly notched at the apex, 1.5-2 mm long. Each silicle contains 2 pale brown, ovoid seeds, about 1.5 mm long.
Flowers: November-January; Fruit: December-February.

Distribution: *South Island. Endemic. Lepidium kirkii* has been recorded from near Alexandra and locations near Ranfurly, Gimmerburn, and Waipiata on the Maniototo Plains in Otago. The only populations known are near Gimmerburn and Alexandra. *Lepidium kirkii* is restricted to low-rainfall areas.

Habitat: Plants grow along the margins of salt pans which form in shallow depressions amongst short grassland.

Conservation: Sizes of extant populations are uncertain and the species is threatened by changes in land use and irrigation. Surveys are underway and attempts will be made to conserve this species. Only a very few plants are in cultivation.

Comments: *Lepidium*: a scale, referring to the shape of the silicle; *kirkii*: after Thomas Kirk (1828-1898), an English botanist who arrived in New Zealand in 1863.

Status: Endangered.

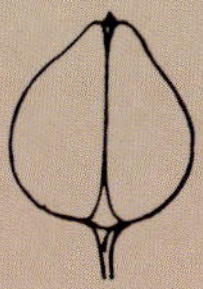

seed

P. Brooke

G. Loh

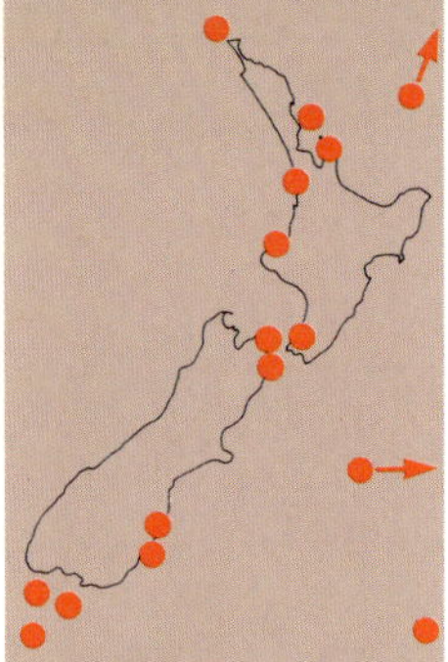

P. Brooke

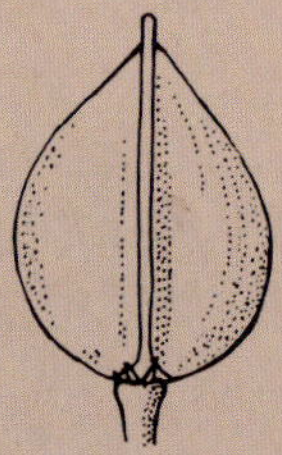

seed

C. Ogle

E. Cameron

Lepidium oleraceum

nau; Cook's scurvy grass

Brassicaceae

Description: Stout, spreading or erect herb up to 50 cm tall with much-branched stems tending to be woody at the base. Lower leaves fleshy, bright green, laminae narrowly oblanceolate to obovate, up to 10 x 3 cm, margins evenly toothed towards the tip, tapering to a broad, flat petiole. Upper leaves similar but somewhat smaller. Flowers small, white, up to 3 mm, arranged in clusters along branchlets or at the branchlet tips. Fruit a flattened silicle, broadly ovate in outline, 3-5 mm long, with truncated base and sharply pointed apex. Each silicle contains 2 brown seeds which are ovoid and 1.5-2 mm long.
Flowers: November-March; Fruit: December-April.

Distribution: *North Island, South Island, Kermadec Islands, Three Kings Islands, Poor Knights Islands, Chatham Islands, Solander Islands, Snares Islands, Antipodes Islands. Also on Lord Howe Island*. Cook's scurvy grass was once more widespread along rocky parts of the New Zealand coast. On the main islands it is now known only from a single North Island site south of Port Waikato and several sites on the south-east coast of South Island. Colonies also remain on some offshore islands and islets near Coromandel, Ocean Beach, New Plymouth, Porirua, Dunedin, Stewart Island, and in the Marlborough Sounds.

Habitat: *Lepidium oleraceum* grows in fertile soils on coastal slopes, disturbed talus, gravel beaches, and rocky shorelines. It is often found near the nesting and roosting sites of sea birds.

Conservation: Browsing animals have undoutedly contributed to the decline of this species. Other factors probably include fungal diseases of cresses and predation by insects. *Lepidium oleraceum* is protected through reserve status of several sites but many sites remain unprotected. It is easily raised from seed, grows vigorously and is occasionally found in private gardens. The foliage of cultivated plants often becomes infested with a white fungus (*Albugo* sp.).

Comments: Few species in New Zealand have declined as dramatically as this one. In 1769 Lt James Cook gathered *Lepidium oleraceum* and celery (*Apium* sp.) by the dinghy-load. Cook's use of *L. oleraceum* as an antiscorbutic or remedy for scurvy (a disease caused by a lack of vitamin C in the diet) led to the common name of Cook's scurvy grass. *L. oleraceum* shows great morphological variation in habit, and leaf size and shape throughout its geographic range. When not in flower it can occasionally be confused with sea rocket and *Senecio lautus* which also grow in coastal habitats and sometimes have similar leaves. The *Senecio* lacks the sharp cress flavour of *L. oleraceum*. *Lepidium*: a scale, referring to the shape of the silicle; *oleraceum*: used as a herb or vegetable.

Status: Vulnerable.

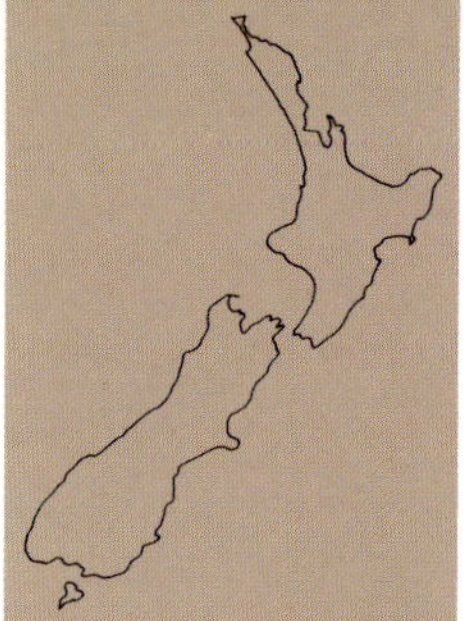

Lepidium obtusatum

Brassicaceae

Description: Leafy herb resembling *Lepidium oleraceum* (p. 82). Lower leaves with broad, flat petioles longer than the laminae, lamina obovate-spathulate, margins toothed, 3-7 cm x 0.5-2 cm. Upper leaves smaller, petioles somewhat shorter or absent altogether. Flowers small, about 1.5 mm, clustered at tips of short, leafy branchlets. Fruit a flattened silicle, broadly ovate, 4.5-5.5 mm long, shallowly notched at the apex, abruptly tapered towards the base. Each silicle contains 2 brown seeds about 2 mm long and broadly ovoid to triangular. Flowering and fruiting: December-July.

Distribution: *No wild populations known. Endemic.* This species has been collected near Auckland and Wellington, in the North Island, but has not been seen since the early part of this century and is probably extinct. It is not known to be in cultivation.

Habitat: Brief notes on herbarium sheets indicate that *Lepidium obtusatum* was collected from rocky coastal headlands.

Comments: *Lepidium*: a scale, referring to the shape of the silicles; *obtusatum*: blunt, possibly referring to the blunt teeth along the margins of the upper stem leaves.

Status: Presumed extinct.

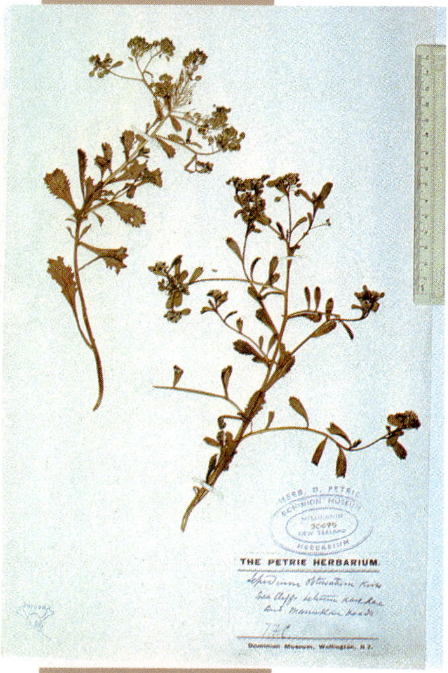

Leptinella nana

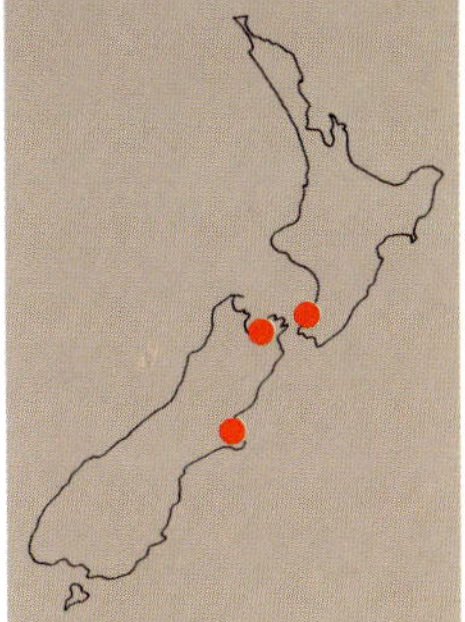

Asteraceae

Description: Very small, creeping herb forming small patches of thin, bright green leaves, distributed along prostrate stems. Each leaf up to 2 cm long, obovate, divided into 6-10 pairs of pinnae. Flower stems up to 1 cm long, terminating in a single, greenish yellow flower head about 2 mm in diameter. Fruit of shallowly wrinkled, greenish brown achenes, about 1 mm x 0.5 mm.
Flowers: April-May and August-October; Fruit: May-June and September-November.

Distribution: *North Island, South Island. Endemic.* Only three small populations of this tiny *Leptinella* are known. They occur near Wellington, in the Rai Valley in Marlborough, and on the Port Hills near Christchurch.

Habitat: *Leptinella nana* is found on bare clay and muddy soils in lowland or coastal areas. The largest population, near Wellington, grows along the top and on the lee side of coastal cliffs on soils which are very dry in autumn. The Rai Valley plants grow on a river bank beneath willow trees and the colony near Christchurch grows in damp ground below steep bluffs.

Conservation: Several hundred specimens of *Leptinella nana* are estimated to grow in the wild. The species is easily overwhelmed by other plants and seems to have little ability to re-establish itself after burial by wind- or water-borne soil. The nearby presence of a small gorse patch poses some threat to the Wellington colony and severe flooding appears to have destroyed some plants in the Rai Valley. The number of plants at the Wellington site has been increased by the devoted activities of a local guardian who has divided some of the existing plants, but attempts to establish *Leptinella nana* in other areas near Wellington have so far been unsuccessful. The population near Christchurch is in a scenic reserve. *Leptinella nana* is easy to propagate but may require constant weeding and attention. A few plants are in cultivation.

Comments: *Leptinella nana* is the smallest of our button daisies. Formerly in the genus *Cotula. Leptinella*: slender; *nana*: dwarf.

Status: Endangered.

J. Miles

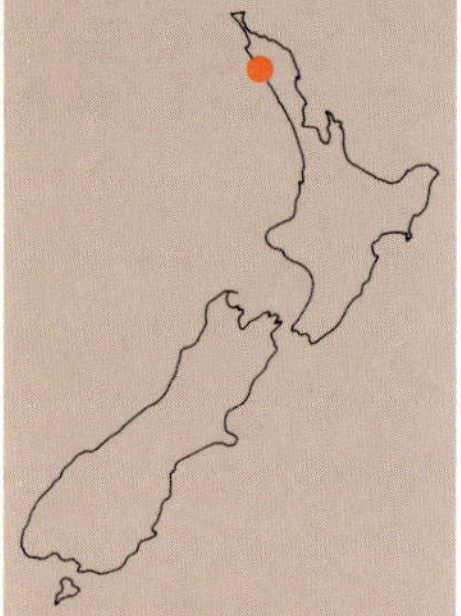

Leptinella rotundata

Asteraceae

Description: Small, creeping, sparingly branched herb forming a loose turf of yellowish green leaves along prostrate stems. Leaves 1-5 x 0.5-1.5 cm, lamina suborbicular, sparsely covered in long hairs, margins toothed, 0.5-1.5 cm long, petiole long. The flower stems are 2-6 cm long and terminate in a single green flower head up to 7 mm in diameter. Each smooth brown achene is up to 1.9 mm long and 1.1 mm wide. Flowers in summer, with long-persisting fruiting heads.

Distribution: *North Island. Endemic. Leptinella rotundata* has been recorded from only two (or possibly three) sites in the northern half of the North Island. Specimens were collected near Waitakere early this century, but the exact location of this site is not known. The only population known at present is on the west coast south of Hokianga.

Habitat: This button daisy grows amongst low vegetation on coastal cliffs.

Conservation: Probably no more than a few dozen plants remain at the one locality, which is within a scenic reserve. Considerable erosion due to heavy seas may threaten this site and there is a possibility that part of the escarpment where the plants are growing may collapse. Nursery propagated stock was replanted into the wild in 1978 but was later destroyed by the intrusion of livestock. *Leptinella rotundata* is easily cultivated and propagated from offsets but is often overwhelmed by other plants. It is sometimes grown in New Zealand often as an attractive ground cover for rock gardens.

Comments: *Leptinella dioica* resembles *L. rotundata*, but has fleshy, hairless, obovate or elliptic leaves. *Leptinella*: slender; *rotundata*: rounded.

Status: Endangered.

D. Given

Scutellaria novae-zelandiae

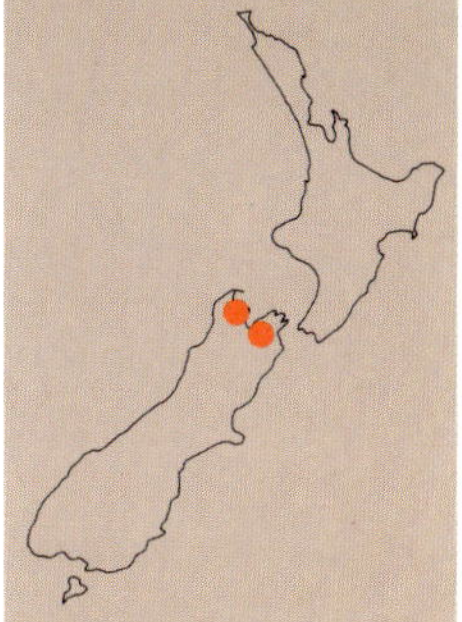

Lamiaceae

Description: Softly woody herb up to 30 cm, stems creeping, branchlets erect or spreading, rectangular in cross section, sometimes covered in short hairs. Leaves small, petioles slender, dark, up to 2 cm long, laminae dark green above, often purplish below, 3-13 x 2-16 mm, shape is very variable, ovate to suborbicular or almost kidney-shaped in outline with few-lobed, toothed, or entire margins. Solitary white flowers in the axils of the upper leaves are clothed in minute hairs, up to 5 mm in diameter. Fruit capsule opens to leave a curious shovel-shaped scale after the tiny nutlets have been shed.
Flowers: December-February; Fruit: January-March.

Distribution: *South Island. Endemic. Scutellaria novae-zelandiae* was collected for the first time from Foxhill on the Waimea Plains in 1840. Specimens have since been occasionally collected from lowland sites near Takaka in north-west Nelson to the Tinline and Pelorus Valleys in Marlborough. The species is currently known from one location near Takaka and several sites in the Maitai and Roding Valleys near Nelson.

Habitat: *Scutellaria novae-zelandiae* has been found on the forest floor in pockets of remnant lowland black beech forest, beneath tall tea-tree scrub, and on river banks, steep cliffs, and among talus blocks.

Conservation: No estimates of the number of plants surviving in the wild have been made. One of the three largest colonies has survived the construction of a dam nearby, but feral goats are a continuing threat. Another colony is within Abel Tasman National Park. The species is easily grown but is very rarely found in cultivation.

Comments: *Scutellaria novae-zelandiae* belongs to the mint family, the Lamiaceae. *Scutellaria*: having a small saucer or scutella; *novae-zelandiae*: of New Zealand.

Status: Endangered.

R. Lamberts

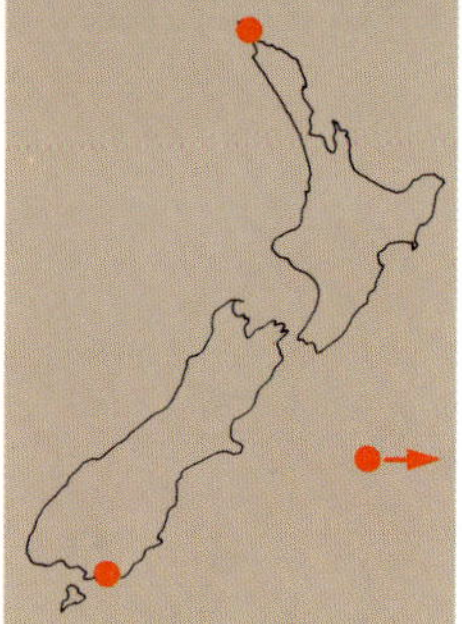

Theleophyton billardierei

Chenopodiaceae

Description: Much-branched, prostrate herb which forms straggly mats, branches yellowish brown somewhat fleshy, up to 40 cm long. Leaves thick, fleshy, green, covered in tiny, glistening papillae and becoming silvery white and somewhat mealy when dry, laminae 5-15 x 3-10 mm, ovate-oblong to obovate or elliptic, margins entire to irregularly toothed. Flowers in clusters of 2-5, tiny and greenish; male flowers in the axils of leaves near the tip of branchlets; solitary female flowers enclosed in small bracts in axils of lower leaves. The reddish brown seeds are 2 mm long and 1.5 mm wide.
Flowers: November-February; Fruit: December-April.

Distribution: *North Island, South Island, Chatham Islands. Also found in Australia and Tasmania.* *Theleophyton billardierei* was once locally distributed on beaches throughout New Zealand. Since 1970 it has been reported from North Cape and Houhora in North Auckland, the south-east of the South Island, and the northern coast of Chatham Island.

Habitat: This is a species of sandy habitats, and is generally found on open beaches and sandy banks close to the high tide mark, where it forms low sandy mounds with only the tips of the branchlets showing. Sea rocket, which is naturalised in New Zealand, occupies identical habitats.

Conservation: The cause of decline in this plant is not known although it is possible that it is very vulnerable to disturbance. None of the few sites where it occurs are reserved and no specimens are known to be in cultivation.

D. Given

Comments: This very distinctive species belongs to the spinach family and is one of a group of plants specially adapted to salty environments. It is the only member of the genus *Theleophyton.* *Theleophyton*: a warty plant; *billardierei*: after Jacques de la Billardiere (1755-1834), a French physician and botanist.

Status: Vulnerable.

D. Given

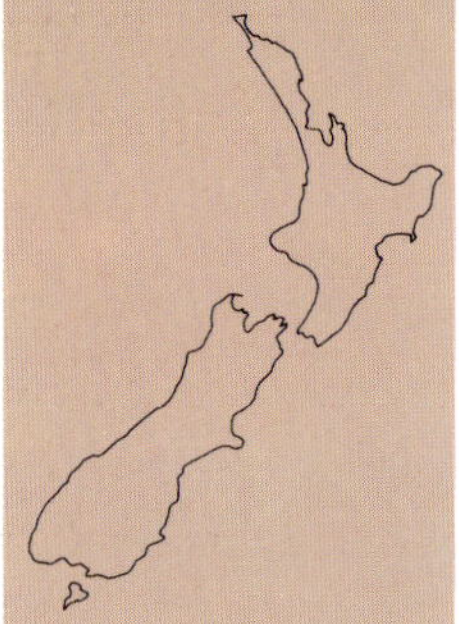

Stellaria elatinoides

Caryophyllaceae

Description: Loosely tufted pale green herb up to 5 cm, with closely branched stems which trail along the ground. Leaves dark green or bluish green, narrowly oblong or oblanceolate, 3-5 mm long, including the winged petiole. Flowers solitary, produced in the leaf axils, lacking petals but with 5 pale green sepals about 3 mm long and with white, membranous margins. Fruit of ovoid capsules about 3 mm long, containing numerous tiny brown seeds 0.6-0.7 mm in diameter. Flowers: October; Fruit: November.

Distribution: *No wild populations known. Endemic.* *Stellaria elatinoides* was first described from specimens collected in the Hawke's Bay region of the North Island in the 1840s and was later collected from three South Island sites between 1890 and 1922: near Ashburton in Canterbury, near Alexandra in Otago, and near Lake Te Anau in Southland. No further reports of this plant have been received in over 60 years and no specimens are known to be in cultivation.

Habitat: *Stellaria elatinoides* was found on dry grassy banks and in bare places in grassland.

Comments: *Stellaria*: star-like, referring to the shape of the flowers; *elatinoides*: resembling *Elatine*, a genus of small aquatic plants.

Status: Presumed extinct.

P. Brooke

Wahlenbergia stricta

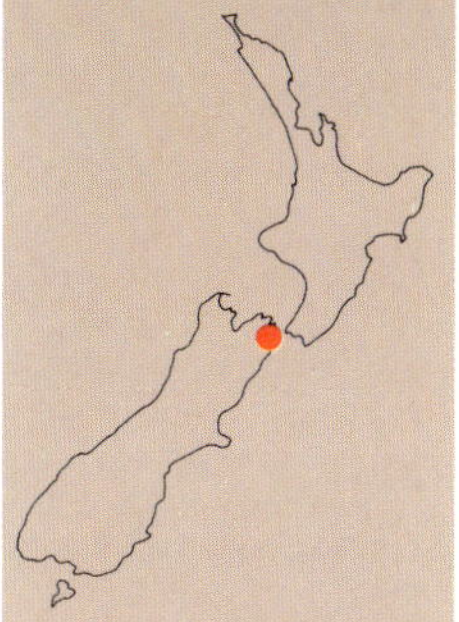

Campanulaceae

Description: Untidy, straggling herb with lax stems up to 60 cm long and, like the leaves and calyx, covered in short, bristly hairs. Leaves 1-2.5 cm x 1-5 mm, narrowly obovate to narrowly lanceolate, and have stiff, bristly hairs on both surfaces, margins wavy, undulate or toothed. The prominent purplish blue bell-shaped flowers are 2-3.5 cm in diameter, and are borne on scabrid to bristly stems. Fruits are scabrid to bristly-globose capsules, each with 5 prominent vertical ribs, 5-7 x 5 mm, containing many minute seeds.
Flowers: November-April; Fruit: December-May.

Distribution: *South Island. Widespread in Australia*. The only population of *Wahlenbergia stricta* found in New Zealand grows along the coast near Blenheim, in Marlborough.

Habitat: Steep coastal slopes amongst bracken and low scrub.

Conservation: Only a few plants survive in an area of a few square metres. The site is reported to be in a high fire-risk area and the surrounding hillsides are being invaded by weeds and *Pinus radiata*. *Wahlenbergia stricta* can be readily grown from seed, but only a few plants are in cultivation, and it may be short-lived.

Comments: Until recently New Zealand specimens were included in *Wahlenbergia trichogyna*. *Wahlenbergia*: after the Swedish botanist Goran Wahlenberg (1780-1851); *stricta*: possibly referring to the erect flower stems.

Status: Endangered.

B. Given

orchids

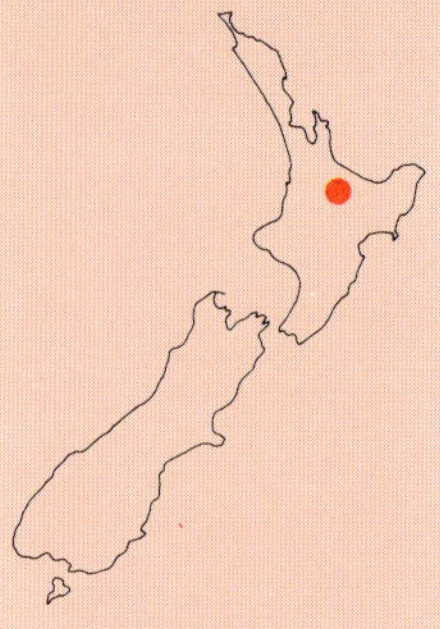

Caleana minor

small duck orchid

Orchidaceae

Description: Very slender orchid with a single narrow, linear leaf 4-10 cm long attached at the base of the plant. The thin, red flower stem is 5-20 cm high and supports 1-7 yellowish green flowers which each resemble a flying duck in profile. The flowers are held "upside down" so that the labellum, which is covered in raised purple glands, is at the top. Below the labellum a long green column forms a cup up to 8 mm deep. Flowers: November-January.

Distribution: *North Island. Common in Australia*. Small duck orchids were recorded from Kaitaia, Rotorua, and Waiotapu in the early 1900s but were not seen again until a population was discovered near Rotorua in 1979.

Habitat: In the past this orchid has been found on dry sandy hills, gumland, clay land, and in open patches of manuka. The only known population grows amongst light scrub on geothermal ground.

Conservation: Only three clumps of small duck orchids have been found. Together they comprise a population of about 12 plants which are reproducing vegetatively by forming new tubers. Although flowers are usually destroyed by insects, a relatively large number of seeds were produced in 1982, aided by artificial pollination. The site is threatened by trampling, clearing, burning, and digging. Attempts to maintain small duck orchids in cultivation have so far been unsuccessful.

Comments: The highly modified flower of this orchid has a hinged labellum which springs back into the column whenever an insect lands on it, thus ensuring that the flower is pollinated by the insect. *Caleana:* after George Caley (1770-1829), an early Australian botanical collector; *minor*: smaller (than *Caleana major*).

Status: Endangered.

D. Blake

Calochilus campestris

copper beard orchid

Orchidaceae

Description: Similar to *Calochilus robertsonii* (p.96) but with a smaller labellum which is only a little larger than the sepals. The base of the labellum is covered with dark blue ridges and has a single dark gland on either side. Its middle section is covered with reddish blue hairs and tapers to a short, hairless ribbon which is clearly visible at the tip of the labellum.
Flowers: October-November.

Distribution: *North Island. Widespread in Australia.* Henry Bleneowe Matthews made several collections of copper beard orchids near Kaimaumau, in North Auckland, in the early 1900s. The species was not seen in New Zealand again until 1986, when it was rediscovered near Kaimaumau.

Habitat: *Calochilus campestris* has been found in grassland, scrub, and along roadsides.

Conservation: Less than 10 plants have been found amongst manuka scrub on private land. The dense scrub appears to be overwhelming some plants, and fire is a potential threat to the whole population. Copper beard orchids are difficult to grow in cultivation and probably depend on an associated fungal partner.

Comments: Although at present listed as *C. campestris,* New Zealand plants probably belong to a different species. *Calochilus*: beautiful lip, referring to the elaborate labellum; *campestris*: of field, referring to the orchid's habitat.

Status: Endangered.

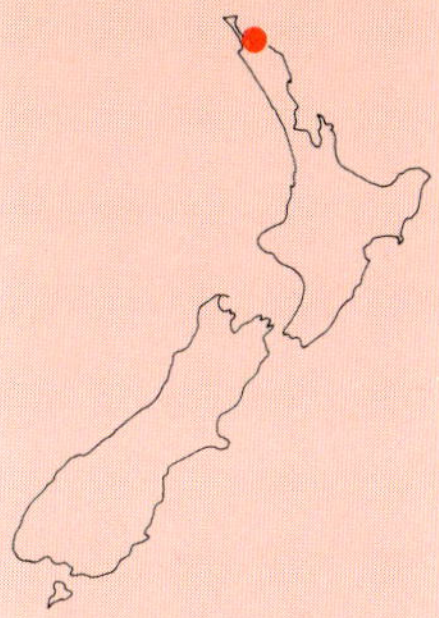

P. Brooke

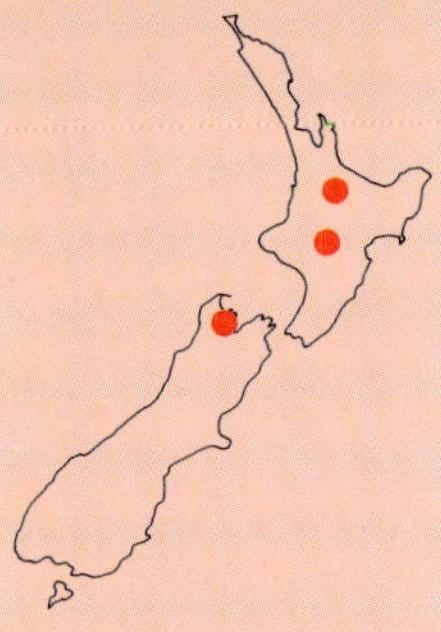

Calochilus robertsonii

red beard orchid

Orchidaceae

Description: Slender to robust orchid with a single, erect, linear- lanceolate channelled leaf and up to 15 flowers on a solitary stem up to 50 cm high. Flowers with greenish red petals and sepals with purplish or reddish brown stripes, a prominent upper sepal which is arched into a hood over the column, and a spectacular labellum which is 20-30 mm long - twice as long as sepals. Labellum base covered with purplish rounded thickenings and has a single, dark bead-like gland on either side, middle section densely bearded with purplish hairs which obscure the short, hairless ribbon at the tip.
Flowers: October-November.

Distribution: *North Island, South Island. Common in Australia*. Past records of red beard orchids are confined to the region between Rotorua and Taupo in the North Island. The species is currently known from sites near Rotorua and in the Kaimanawa Range in the central North Island, and near Kaiteriteri in the north of the South Island.

Habitat: *Calochilus robertsonii* is usually found in open grassland and scrub, often on poor soils and geothermal ground.

Conservation: The largest population in New Zealand occurs amongst manuka scrub at a race-course. Its survival there depends on careful management with the site being left relatively undisturbed - not mown when the orchid is flowering, nor being excessively fertilised or oversown with grass seed. Very few specimens are in cultivation.

Comments: *Calochilus robertsonii* is one of three species of beard orchid found in New Zealand. *Calochilus*: beautiful lip, referring to the elaborate labellum; *robertsonii*: after John Robertson (1803-1862), an early Victorian naturalist.

Status: Vulnerable.

J. Barran

Chiloglottis formicifera

ant orchid

Orchidaceae

Description: Slender orchid bearing 1 small green flower with brown or purplish markings on a stem up to 12 cm high. Labellum is almost diamond-shaped and has 2 groups of small glossy thickenings at the tip and base which resemble insects; the largest is double-headed and resembles an ant. Two sessile, broadly ovate or oblong leaves about 5 cm long are attached at the base of the stem.
Flowers: August-September.

Distribution: *No longer found in New Zealand. Relatively abundant in parts of south-east Australia.* The only specimens of *Chiloglottis formicifera* found in New Zealand were collected between 1900 and 1903 by Henry Matthews, who discovered the species near Lake Tongonge, near Kaitaia, in the far north of the North Island.

Habitat: *Chiloglottis formicifera* probably grew amongst gumland scrub.

Conservation: *Chiloglottis formicifera,* like several other Australian orchids, may have made temporary landfalls in New Zealand but failed to become established here. Profound habitat disturbance, associated with the draining of Lake Tongonge and the conversion of nearby scrub for farming, may have contributed to the disappearance of this immigrant. Only a few ant orchids are known to be in cultivation.

Comments: The ant orchid is easily distinguished from *Chiloglottis cornuta* which is common throughout New Zealand, by its small flower on a long stem. The shape, colour, and ornamentation of the labellum are also distinct. *Chiloglottis*: lip tongue, alluding to the tongue-like labellum; *formicifera*: ant-bearing, referring to the ant-like thickening on the labellum.

Status: Presumed extinct.

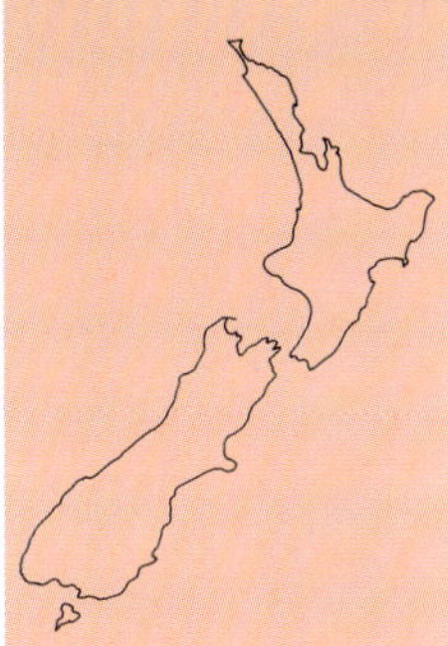

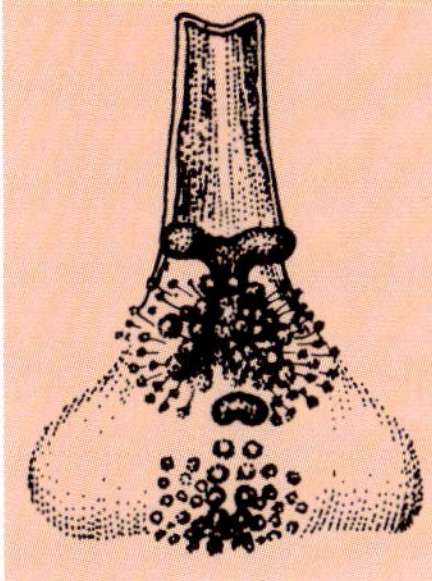

flower

P. Brooke

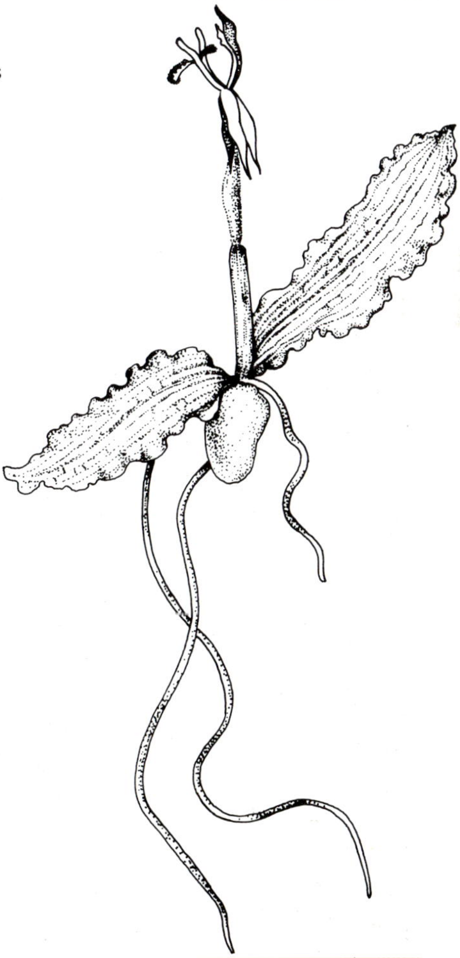

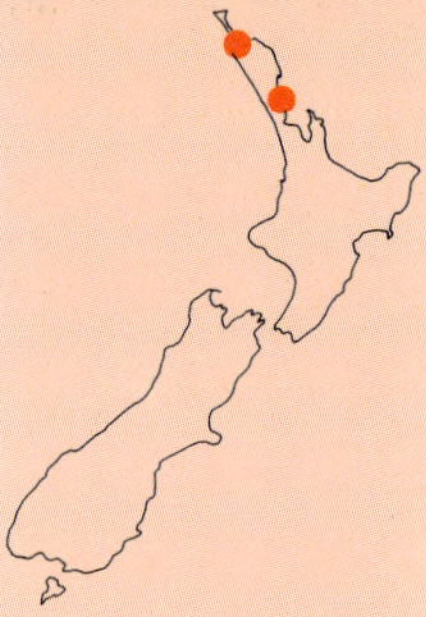

Corybas unguiculatus

small helmet orchid

Orchidaceae

Description: A small, single-leaved orchid 1-3 cm high. The flat, ovate to heart-shaped leaf is 1-3 cm long, green above and reddish below. A single deep reddish purple flower 8-10 mm long is held directly above the leaf. Its upper sepal forms a narrow arching claw which partially encloses the horizontal tubular labellum, so that the flower resembles a duck's head in profile. Flattened, thread-like lateral sepals and petals spread out below and alongside the labellum.
Flowers: June-October.

Distribution: *North Island. Also occurs in Australia.* *Corybas unguiculatus* was recorded from a few localities near Awanui, Kaitaia, Kaiaka, and Aponga in North Auckland early in the century. It is presently known from a few sites between Kaimaumau and Warkworth.

Habitat: Colonies of this orchid have been found in secondary forest and scrub.

Conservation: Few deliberate measures have been taken to preserve the small helmet orchid. One of the largest populations grows on private land and another occurs within a scenic reserve. No specimens are known to be in cultivation.

Comments: *Corybas*: dancing priest, after the worshippers of the Phrygian goddess Rhea Cybelle; *unguiculatus*: fingernail, alluding to the shape of the hooded flower.

Status: Endangered.

D. Climie

D. Climie

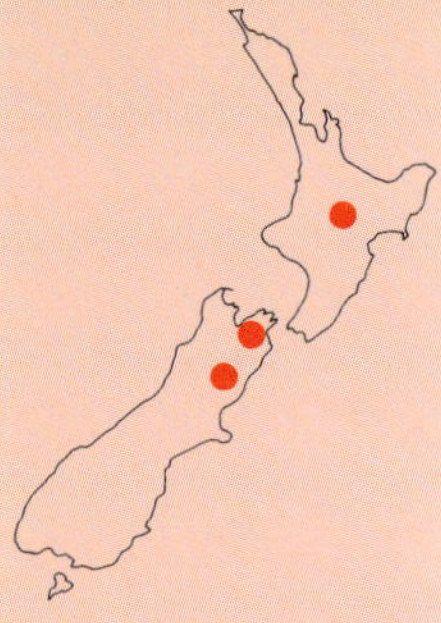

Chiloglottis gunnii

bird orchid

Orchidaceae

Description: Small, robust orchid bearing a solitary flower on a stem usually 4-8 cm high. Flower dark green to reddish brown opening widely to reach 2-3 cm in diameter. The broadly ovate labellum has a long, stalked, purplish thickening at the base, another small, sessile thickening in the centre, and from 1 to 4 smaller purplish glands on either side of the centre line. The 2 green ovate to ovate-oblong leaves are 2-7 cm long and spread opposite each other at ground level.
Flowers: November-January.

Distribution: *North Island, South Island. Common in Australia. Chiloglottis gunnii* was first discovered in New Zealand near Hanmer, Canterbury, in 1981. A second population was found in the same year on the Richmond Range in Marlborough. The first North Island population was discovered in Kaingaroa State Forest in 1986.

Habitat: Near Hanmer the bird orchid grows on the edge of a stand of European larch; on the Richmond Range it grows on the floor of a native beech forest, and the North Island population grows in exotic pine forest.

Conservation: The Hanmer colony is thriving and increasing in numbers. Two of the sites where *Chiloglottis gunnii* is found are within forest parks and a few bird orchids are in cultivation. The North Island site is now protected.

Comments: This Australian immigrant seems to be one of our most recent additions to the native orchid flora of New Zealand. It is similar to *Chiloglottis cornuta*, one of the most widespread orchids in New Zealand. *C. cornuta*, however, is smaller overall and has a green or yellowish green flower on a shorter stem. Its labellum has several large, rounded, green or purplish thickenings. *Chiloglottis*: lip tongue, alluding to the tongue-like labellum; *gunnii*: after Ronald Gunn (1808-1881), an early Australian collector of plants and animals.

Status: Vulnerable.

G. Bawden

Cryptostylis subulata

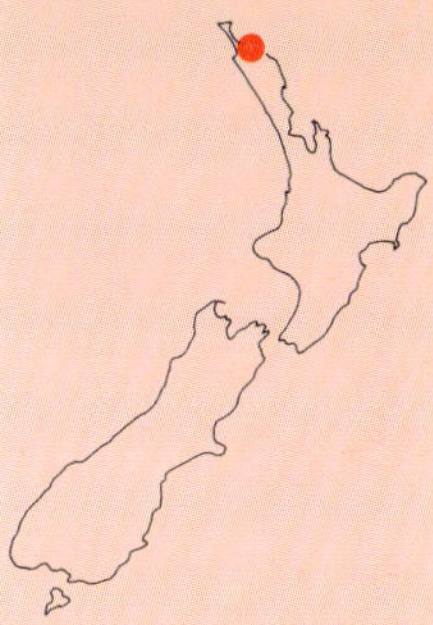

duckbill orchid, tongue orchid

Orchidaceae

Description: Very tall orchid bearing 20 or more flowers on a stem up to 130 cm high. The 1-3 light green leaves with narrow lanceolate laminae are up to 20 x 2 cm wide and taper to narrow petioles nearly as long as the laminae. Flowers are 1.8-2.5 mm long and held "upside down", so that the labellum is at the top. The oblong labellum is reddish plum at the tip fading to yellowish green towards the base, margins are reflexed along the mid-portion to reveal a dark purplish black ridge extending along the mid line almost to the tip. Yellowish green sepals and petals, which are rolled in on themselves to form tight cylinders, radiate out from the base of the flower.
Flowers: December-January.

Distribution: *North Island. Relatively common in Australia*. This spectacular orchid was found in New Zealand for the first time on the Kaimaumau-Motutangi wetlands of North Auckland in 1975. Since then several other populations have been found on nearby Karikari Peninsula.

Habitat: The duckbill orchid is most common in seasonally wet and low-lying areas of peat bogs where it grows amongst sedges, manuka, and ferns.

Conservation: *Cryptostylis subulata* is thriving in New Zealand and appears to be extending its range. It occurs in two proposed scientific reserves - near Lake Ohia and on the Kaimaumau wetlands. Duckbill orchids are easy to maintain in cultivation and are becoming widely cultivated in the north of the North Island. New plants can be produced vegetatively and by hand-pollination.

Comments: This is one of only two New Zealand ground orchids which do not die-back to ground level after flowering but remain leafy throughout the year. It may have arrived naturally from Australia and become established within the last 20 years. The flower is believed to mimic a female ichneumon wasp, thus attracting male wasps which pollinate the flower by attempting to "mate" with it. This wasp is also present in New Zealand and appears to be pollinating wild populations of duckbill orchids. *Cryptostylis*: hidden style, referring to the way in which the column is almost concealed by the colourful labellum; *subulata*: awl-shaped, referring to the sepals and petals.

Status: Vulnerable.

E. Cameron

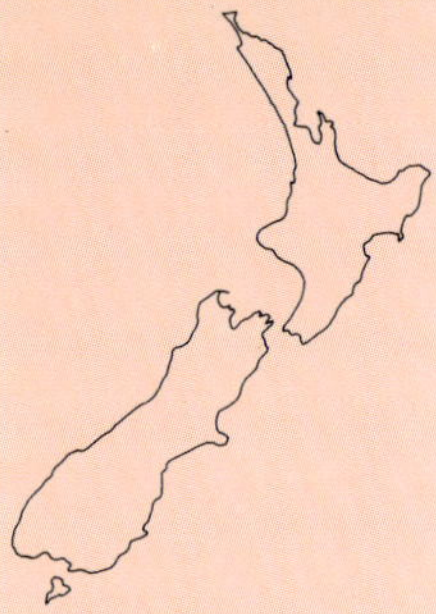

Pterostylis nutans

nodding greenhood

Orchidaceae

Description: Slender orchid readily recognised by the sharp forward and downward incline of the solitary, green, hood-shaped flower which is 1.5-2 cm long and supported on a stem 5-20 cm or sometimes up to 30 cm high. Numerous green leaves form a rosette at the base of the flower stem, and have broadly elliptic to oblong-ovate laminae which are 1.5-3 x 1-2 cm. Flowers: August-September.

Distribution: *No longer found in New Zealand, but common in some parts of Australia*. Records of nodding greenhoods are confined to the North Auckland region of the North Island. The species was collected around Kaitaia between 1910 and 1920 and was last recorded from Whangaparoa Peninsula a few years later.

Habitat: *Pterostylis nutans* was usually found amongst scrub.

Conservation: Habitat disturbance associated with farm development and other activities involving the removal of scrub may have led to the extinction of this species in New Zealand, or prevented the establishment of temporary immigrants from Australia. Nodding greenhoods are very rarely found in cultivation.

Comments: A characteristic feature of *Pterostylis* orchids is the concave hood of the flower. The labellum is attached by a hinge and flicks back into the hood when touched, thereby assisting pollination by insects. *Pterostylis*: winged style, referring to the prominent wings on the column; *nutans*: nodding, referring to its distinctive inclined flower.

Status: Presumed extinct.

R. Lamberts

Thelymitra matthewsii

spiral sun orchid

Orchidaceae

Description: A slender orchid with stems up to 15 cm high which bear 1-2 purplish blue flowers with dark veins. The column of each flower supports 2 fleshy, bright yellow arms. Each plant has a single flexuous or spirally twisted linear leaf, 3-6 cm long.
Flowers: August-November.

Distribution: *North Island. Rarely found in Australia.* Most collections of this species were made early this century, chiefly from low hills between Kaitaia and Awanui and the coast of North Auckland, in the North Island. It has just been rediscovered near the northern tip of New Zealand.

Habitat: *Thelymitra matthewsii* was found in swamps and bogs.

Conservation: Most collected specimens may represent a few immigrants from Australia which never became permanently established in New Zealand. Draining and reclamation of the spiral sun orchid's wetland habitat may have contributed to its disappearance. No spiral sun orchids are known to be in cultivation (at least in New Zealand).

Comments: *Thelymitra matthewsii* is also rare in Australia. *Thelymitra*: female headband, referring to the hooded shape of the column; *matthewsii*: after Henry Matthews (1861-1934), an early collector of orchids in New Zealand, especially in North Auckland.

Status: Endangered.

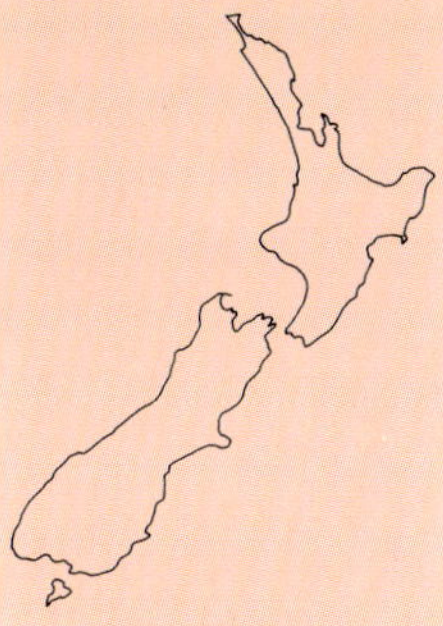

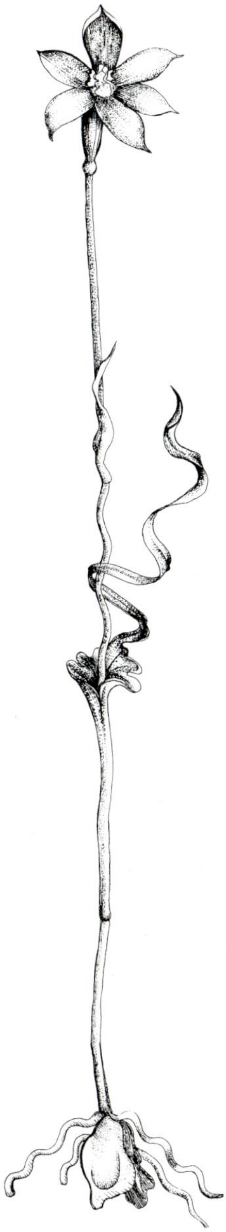

P. Brooke

grasses and grass-like herbs

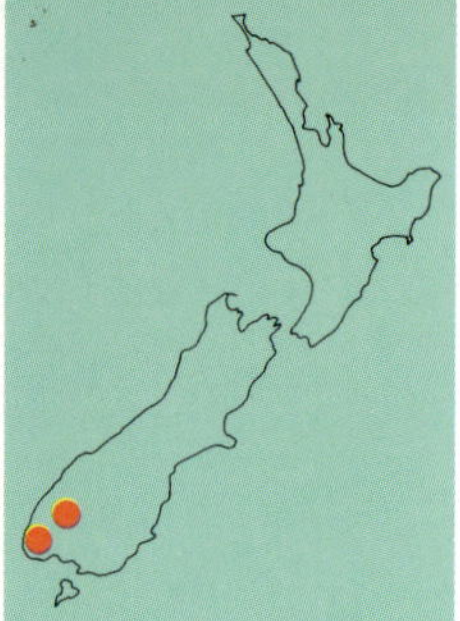

Chionochloa spiralis

Poaceae

Description: Perennial tussock-forming grass. Leaves narrow, up to 60 cm x about 1 mm, tips sharply pointed. It is readily recognised by the flattened, straw-coloured sheaths at the base of each leaf which coil inwards and readily break transversely when dry. Branched panicles bear numerous spikelets each composed of 4-6 small flowers. The lemma surrounding each flower is hairy on the outer surface and about 2 cm long, including its long awn which is flattened and twisted at the base. Flowering and fruiting periods not known.

Distribution: *South Island. Endemic. Chionochloa spiralis* has been found only in Takahe Valley and near Lake Monk in Fiordland.

Habitat: All the plants known in the wild grow on limestone bluffs above or near the tree line.

Conservation: Only a few hundred plants have been found at the two localities, both of which are in Fiordland National Park. One site is remote and access to the other is by permit only, and most of the plants are inaccessible to deer. However, the populations are not large. Very few specimens of *Chionochloa spiralis* are in cultivation. Plants can be readily propagated from tillers and large tussocks are easily divided into plantlets.

Comments: *Chionochloa*: snowgrass; *spiralis*: spiraled, referring to the coiling of the leaf sheath on drying.

Status: Vulnerable.

D. John

Cortaderia turbaria

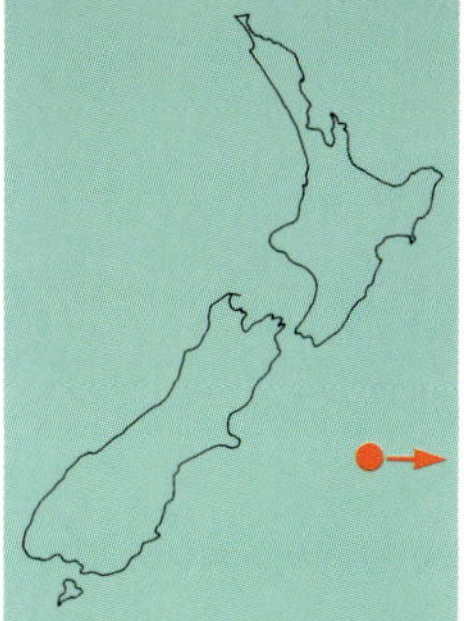

Chatham Island toitoi

Poaceae

Description: Perennial tall, tussock-forming grass up to 2 m. Leaves narrow, up to 1 cm wide and strongly scabrid. It is readily recognised by the long hairs found on sheaths, leaves, panicles, and pedicels. Branched panicles are up to 30 cm and very showy. The bracts (glumes) subtending the flower are about 20 mm long and the lemma up to 1 mm. Flowers are perfect, unlike mainland species. Fruit a caryopsis 3 mm long. Flowering and fruiting: mid to late summer.

Distribution: *Chatham Islands. Endemic. Cortaderia turbaria* occurs as scattered populations on Chatham Island especially in incised streams. It is also recorded from Pitt Island.

Habitat: Plants occur in peaty swamps, and along lake and stream margins.

Conservation: Most populations consist of very few plants, and few are known to have more than 50 plants. However, isolated plants can set seed. The species has been lost at several sites where it was recorded in the 1960s. Very few plants are in cultivation and attempts at propagation have met with varied success.

Comments: *Cortaderia:* cutting, referring to the leaf margin. *Turbaria*: peat, referring to habitat.

Status: Vulnerable.

D. Given

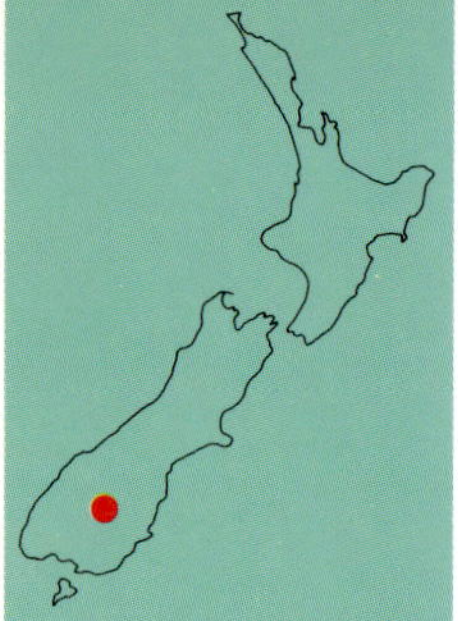

Simplicia laxa

Poaceae

Description: Soft perennial grass, stems weak, decumbent, up to 50 cm long, elongated internodes 5-6. Laminae flat, narrow 6-12 cm x 1.5-3 mm. Sheaths at the base of lower leaves darkly coloured and sometimes clad in minute, soft hairs. Open sparingly branched panicles bear sharply pointed, lanceolate-shaped spikelets towards the tips of branchlets. Each spikelet encloses a single flower and is about 3 mm long. The bracts (glumes) subtending the flower are about 1 mm long and the slightly keeled lemma above is 2.5-3 mm long, ovate-lanceolate, and covered in minute hairs. Fruit a laterally compressed caryopsis about 1.5 mm long.
Flowers: January-February; Fruit: January-February.

Distribution: *South Island. Endemic. Simplicia laxa* was collected in the Ruamahanga Valley in Wairarapa, in the south of the North Island, and from several sites in eastern Otago, in the South Island, in the late 1800s but was not seen again until it was rediscovered on the Old Man Range, in Otago, in 1969.

Habitat: This grass appears to prefer shaded or sheltered sites at the bases of rocks, on bluffs, or in gorges from near sea level to 1000 m.

Conservation: The site where this species was rediscovered is not reserved and the present extent of the population is not known. The species could easily be overlooked and may occur in other localities but was not located during recent surveys of protected natural areas. It is not known to be in cultivation.

Comments: *Simplicia* closely resembles the genus *Poa*, but, unlike *Poa*, has only one flower in each spikelet. *Simplicia*: simple, not compound, referring to the solitary, single-flowered spikelets; *laxa*: loose, referring to the flaccid, decumbent stems.

Status: Endangered.

B. McCulloch

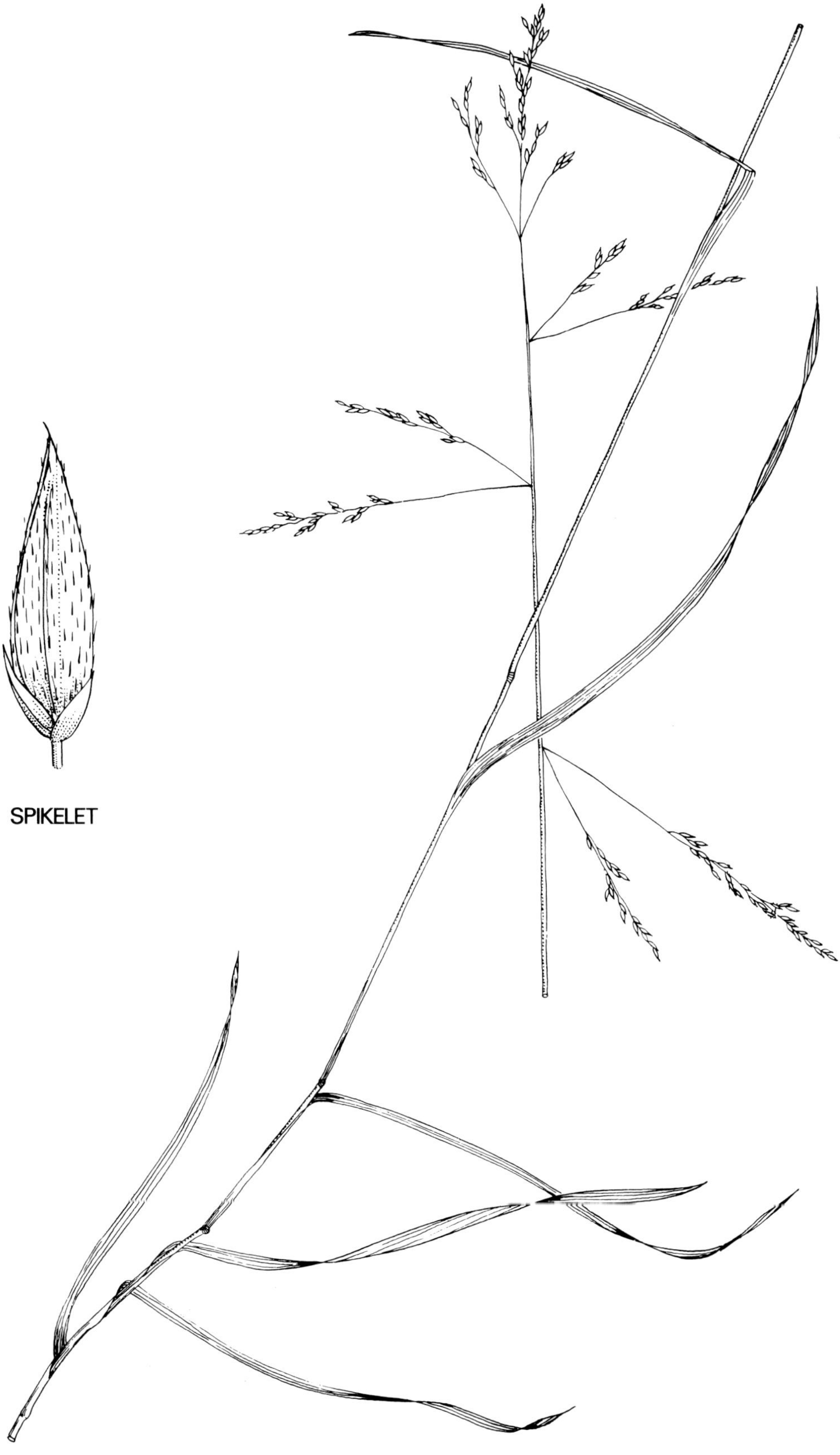

SPIKELET

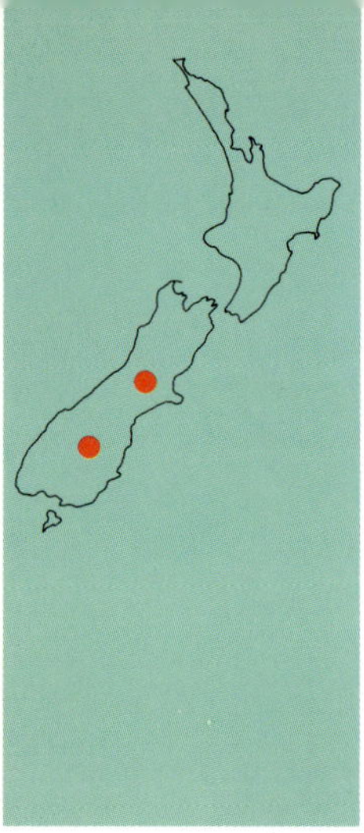

P. Brooke

Triglochin palustre

marsh arrow-grass

Juncaginaceae

Description: Slender, grass-like herb. Leaves in tufts, erect, minutely ridged, linear, 5-50 cm x 0.5 mm, are slightly channelled or semi-circular in cross section. Small greenish or reddish flowers, each about 3 mm long, are borne on a leafless, unbranched stem which elongates up to 40 cm as the fruit develops. The fruit itself is narrowly-linear, 8-10 mm long by 1 mm wide, and is composed of 3 follicles which lack keels. The follicles separate at their very narrow, pointed bases but remain attached at the tip of the central axis, thus resembling the ribs of an opening umbrella.
Flowers: December; Fruit: December-March.

Distribution: *South Island. Widespread in the Northern Hemisphere and also recorded from South America.* Marsh arrow-grass was collected from five inland South Island localities in Canterbury and Otago in the early 1900s but between 1935 and 1955 it was collected only from sites near Lake Tekapo. Three populations known to survive in the wild occur near Lake Coleridge and Lake Clearwater in Canterbury, and near Clyde in Central Otago.

Habitat: Marsh arrow-grass appears to prefer moist soil and muddy to gravelly sites which are often inundated.

Conservation: Large populations of several hundred plants have been found at two of the three known sites. No specific measures have been taken to protect or reserve these sites and no specimens are known to be in cultivation.

Comments: *Triglochin*: three-pointed, referring to the pointed bases of the three follicles; *palustre*: growing in marshes.

Status: Vulnerable

sedges and rush-like herbs

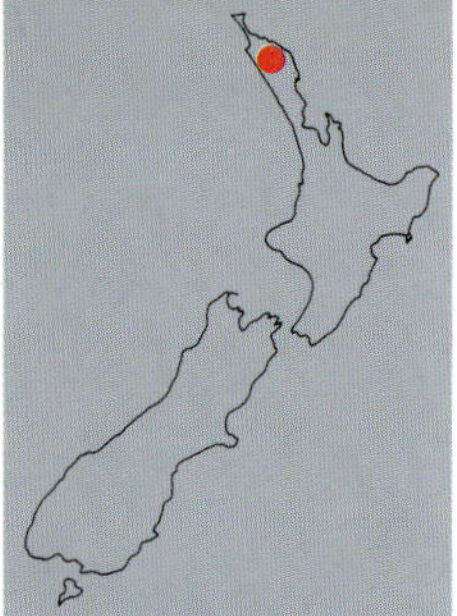

Baumea complanata

Cyperaceae

Description: Leafy, rush-like, tufted sedge. Leaves flat, striated, and sharply pointed, 50-90 cm x 3-15 mm. Stems are rounded or somewhat flattened in cross section, up to 150 cm long. Erect panicles up to 50 cm long are produced near the top of the stems. Each panicle of numerous light reddish brown spikelets about 5 mm long, which each bear 2 or 3 small, inconspicuous flowers. The ripe nuts are triangular in cross section and reddish brown. They are about 4 x 2 mm and taper to a long scabrid beak at the top. Flowering and fruiting periods not known.

Distribution: *North Island. Endemic. Baumea complanata* has been recorded from about 10 sites scattered between Kaikohe (North Auckland) and the Waikato lowlands, but by the early part of this century it had disappeared from more than half of these sites. The only remaining populations known at present occur in the Ngawha basin and at Taheke, both near Kaikohe.

Habitat: The usual habitat of this sedge is damp to seasonally dry areas amongst manuka scrub. It appears to prefer sites near lakes and streams and can also grow in shallow running water.

Conservation: The few remaining populations in the Ngawha basin total some hundreds of plants. The basin has been test drilled to assess the feasibility of geothermal power development; and development of the area for tourism has also been proposed. One of the largest known populations is within a scientific reserve and another colony is protected under a conservation covenant. *B. complanata* is palatable to cattle and seems to soon succumb to repeated browsing, as a favoured food plant of livestock in the nutrient-deficient scrubland where it is found.

Plants taken from the wild have been successfully transplanted in cultivation (sometimes with difficulty), and the establishment of new wild colonies by transplanting in suitable sites has been suggested. Seed germination trials have shown that seed must be incubated for about 4 years before germination occurs.

Comments: Formerly regarded as a subgenus of the genus *Cladium, Baumea* is now recognised as a distinct genus. The flattened rather than cylindrical leaves of this species are more like those of the genus *Machaerina* and distinguish *Baumea complanata* from other species of *Baumea. Baumea*: after Baume, a place in the Philippine Islands; *complanata*: flattened, referring to the flat leaves.

Status: Endangered.

D. Given

Carex chathamica

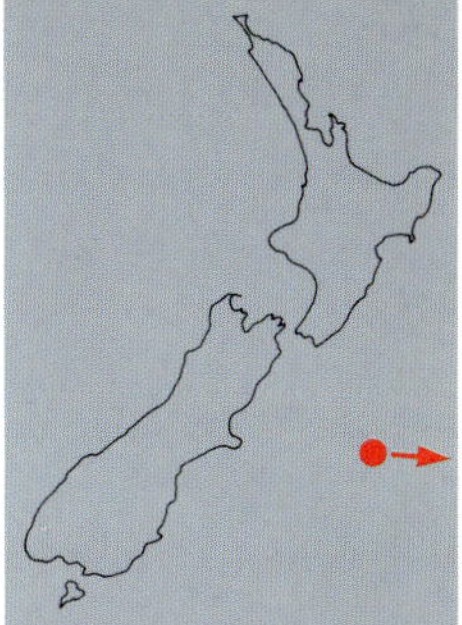

Cyperaceae

Description: Stout, pale green sedge tussock. Leaves rigid, grass-like, shaped like a flattened "M" in cross section, margins finely toothed, up to 35 cm x 8 mm. Flower stems triangular in cross section with tiny flowers clustered together in 6-8 light brown spikes. The uppermost 2-4 sessile, slender spikes are composed entirely of male flowers. Lower spikes up to 10 x 1.5 cm and consist predominantly of female flowers. Each female flower enclosed in a utricle 3-4.5 mm long, subtended by a much longer light brown or purplish red glume with a greenish midrib (see p. 116). Pale greyish brown, oblong-obovoid nuts are triangular in cross section and about 2 mm long.
Flowering and fruiting periods not known.

Distribution: *Chatham Islands. Endemic. Carex chathamica* has been recorded from several sites on Chatham Island and also occurs on Pitt Island.

Habitat: *Carex chathamica* grows in wet clearings, along lake and pond margins, and in swamps.

Conservation: Like the other large sedges, *Carex chathamica* is heavily grazed whenever it is accessible to stock. Continued grazing and modification of habitat, especially on peat domes, threatens its survival in the wild. Few conservation measures have been undertaken for this species. It occurs in one reserve on Chatham Island but further protection of the plant's habitat is required. No specimens are known to be in cultivation.

Comments: *Carex chathamica* is similar in general appearance to *C. ventosa, C. trifida,* and *C. ternaria*, the three other large sedge tussocks found on the Chatham Islands, but can be distinguished from these species by differences in the inflorescence. The female spikes of *C. chathamica* and *C. trifida* are usually greater than 1 cm in diameter, whereas they are narrower in *C. ventosa* and *C. ternaria*. The lower spikes of *C. ternaria* arise in pairs or groups of three from the bases of the leaf-like bracts of the inflorescence, whereas the lower spikes of *C. chathamica* and *C. ventosa* are solitary. *C. chathamica* is usually smaller than *C. trifida* and has smooth utricles, whereas the utricles of *C. trifida* and *C. ventosa* are distinctly ridged. *Carex:* cutting, from an old Greek name for rush or sedge; *chathamica*: from the Chatham Islands.

Status: Vulnerable.

D. John

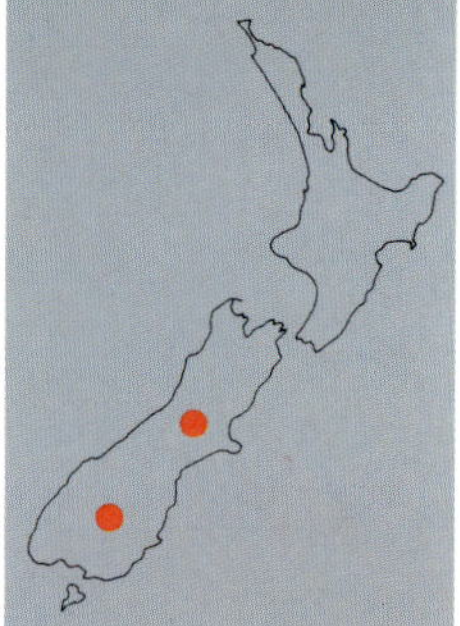

Carex inopinata

Cyperaceae

Description: Rather inconspicuous, leafy, mat-forming sedge with long, creeping rhizomes about 0.5 mm in diameter and grass-like leaves up to 25 cm long x 2 mm wide. Flower stems up to 4 cm are enclosed by brown leaf sheaths. Each flower stem bears 2-3 shortly stalked lower spikes and a very slender upper spike on a stalk up to 6 cm or more long. The slender, upper spike usually consists entirely of male flowers; the 2 lower spikes of 3-7 female flowers. Each female flower has 3 stigmas and is enclosed in a distinctive utricle which is dark reddish purple towards the top, about 6-7 mm long, and has a very narrow beak cleft at the top and about 3 mm long (see p. 116). Oblong nuts enclosed within the utricles are triangular in cross section, greyish brown, and about 2 mm long. Flowers: October-January; Fruit: November-April.

Distribution: *South Island. Endemic. Carex inopinata* is known only from a collection made near Bendigo in North Otago in 1974, and a very small population which grows at Castle Hill in Canterbury.

Habitat: At Castle Hill this sedge is found beneath matagouri and *Coprosma* bushes at the base of limestone outcrops.

Conservation: Nine wild plants have been found at Castle Hill within an area of about 60 m^2. Five out of 6 nursery-raised plants transplanted back at the site in 1980 have survived and another 6 nursery-raised plants have recently been transferred to the same area. The Castle Hill site is being grazed by sheep at present but is included within a proposed reserve. The Otago population may have been relocated but specimens collected require critical identification from fruit.

Carex inopinata is easily grown from offsets and by division of the plant mat. A few specimens are under cultivation in private and botanic gardens.

Comments: This is not a particularly conspicuous member of the genus. When not in flower it is difficult to distinguish from *Carex wakatipu* which is common in snow-tussock grassland on the eastern side of the South Island. *Carex*: cutting, from an old Greek name for rush or sedge; *inopinata*: unexpected, probably referring to the unexpected discovery of the species in this country, as it is the only species in its group known to occur in New Zealand.

Status: Endangered.

P. Brooke

Carex ventosa

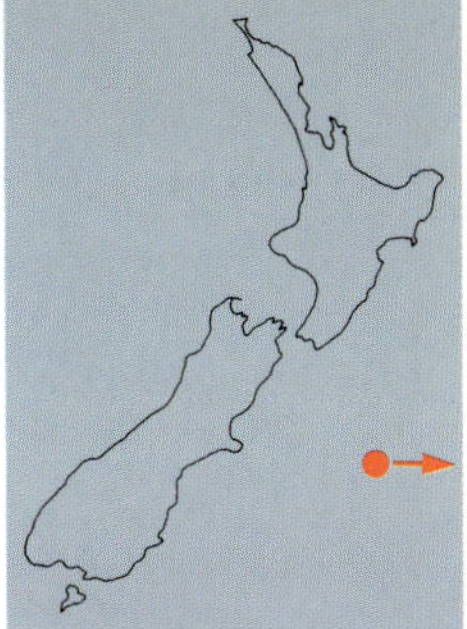

Cyperaceae

Description: A very leafy sedge tussock. Leaves shaped like a flattened "M" in cross section, up to 80 cm x 5-11 mm, but usually much shorter. Flower stems, triangular in cross section, up to 80 cm long, with about 7 spikes of minute flowers. The uppermost 1-2 spikes of male flowers and the remaining mainly of female flowers. Individual female spikes light greenish brown and up to 9.5 cm x 8 mm wide. The distinctly ridged and beaked utricle enclosing each female flower is 4.5-5 mm long and is subtended by a glume which is colourless and transparent apart from minute reddish brown streaks and a greenish midrib. The glumes are shorter than, or approximately the same length as the utricles (see p. 116). The dark brown, oblong nuts enclosed within the utricles are triangular in cross section and about 2 mm long.
Flowering and fruiting periods not known.

Distribution: *Chatham Islands. Endemic. Carex ventosa* is found on Chatham Island, Pitt Island, Mangere Island, and South East Island.

Habitat: This species grows in open coastal forest, on coastal cliffs, and in bush surrounding the burrows of sea birds. It is found in well drained sites which are unsuitable for *Carex chathamica*.

Conservation: Although several thousand plants of this species may remain in the wild, individual populations on Chatham Island itself are usually small. The species was probably once far more widespread on Chatham Island but grazing pressure and the destruction of coastal forest seems to have led to its decline. Numbers could decline still further if peat is mined on Chatham Island and grazing pressure increases at some of the sites where it now grows.

Carex ventosa is found in at least one reserve on Chatham Island, on some island reserves of the Chatham group, and also occurs in several proposed reserves. It should persist on inaccessible sites and in areas of securely fenced coastal forest but needs further reservation on the main island. Only very few are in cultivation.

R. Chappell

Comments: *Carex*: cutting, from an old Greek name for rush or sedge; *ventosa*: swaying in the wind or growing in exposed situations.

Status: Vulnerable.

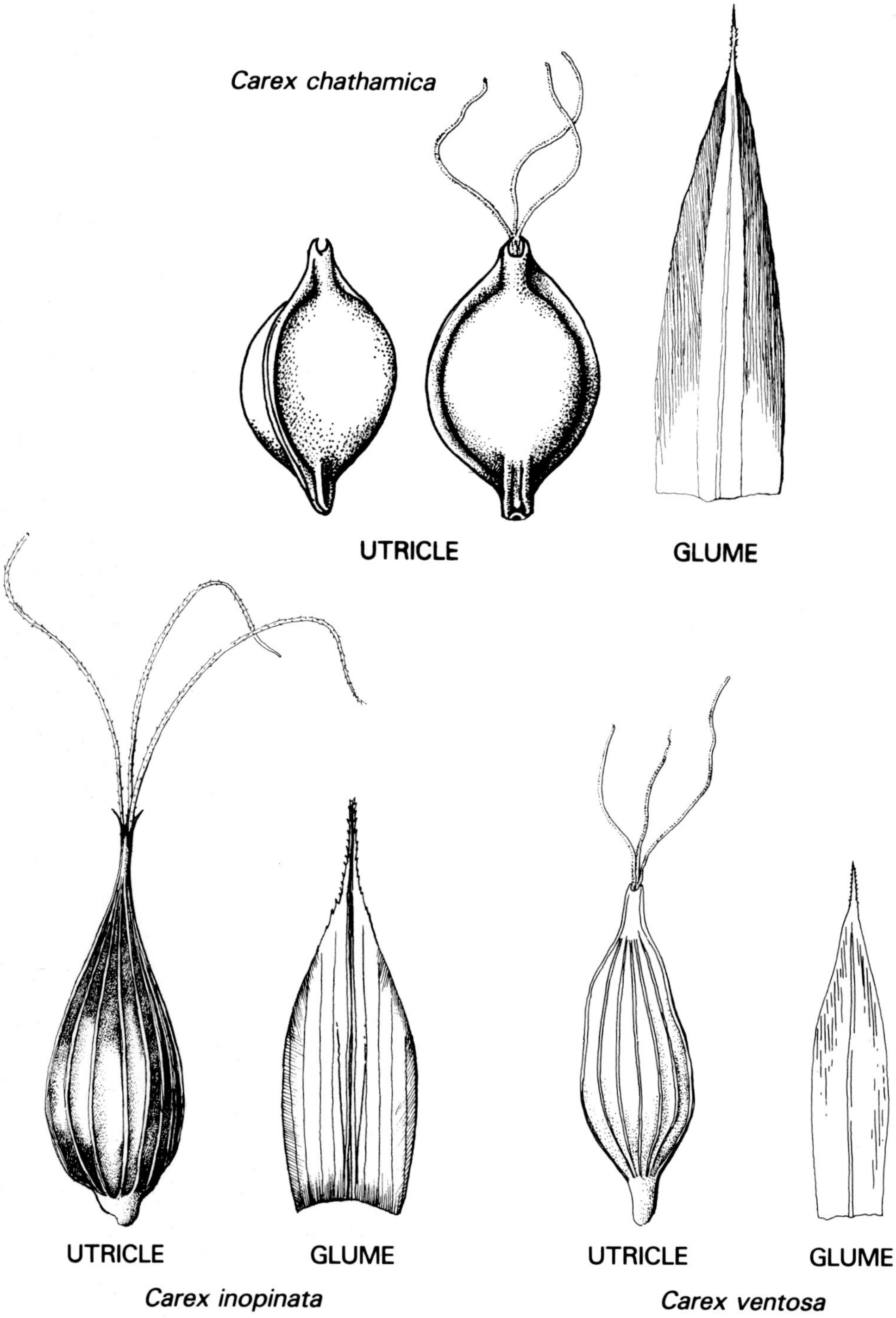

P. Brooke

Comparison of the utricles and glumes of Carex chathamica, C. inopinata, *and* C. ventosa.

Eleocharis neozelandica

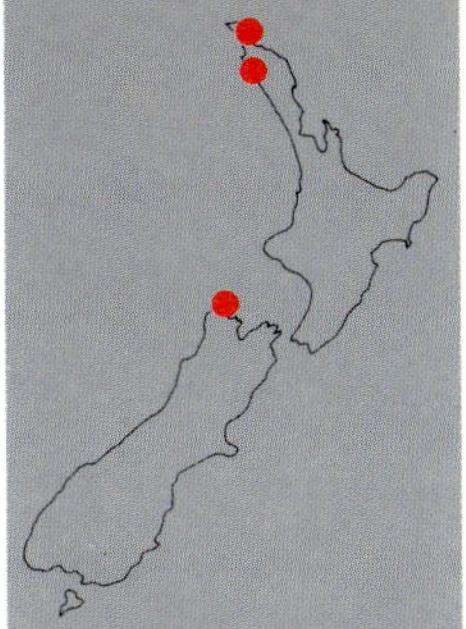

Cyperaceae

Description: Small, leafless sedge with tufts of rigid flowering stems which are usually 3-8 cm x 0.5 mm wide, stems arise from long, dark brown creeping rhizomes about 1 mm in diameter. Each flower stem is enveloped at the base by two membranous leaf sheaths, the lowermost of which is purplish brown. At the stem tip is a solitary, broadly ovoid spikelet, 5-8 x 1-6 mm, which bears 4-10 minute flowers. Flowers each subtended by a glume 2.5-3.5 mm long, consisting of only 2 or 3 stamens and an ovary with 2 style branches. The shiny, golden-brown nuts are rather asymmetrically obovoid and 1.5-2 x about 1 mm. Flowers: November-December; Fruit: January-February.

Distribution: *North Island, South Island. Endemic.* Since the early 1900s *Eleocharis neozelandica* has been recorded from several North Island sites north of Hokianga Harbour in North Auckland, and along the western coast of Wellington Province. In the last two decades, however, it has only been found at four sites north of Kaitaia, and near Hokianga Harbour, Levin, and Foxton. Recent searches have failed to find any trace of the Levin and Foxton populations. South Island specimens of *E. neozelandica* were collected in the early part of this century from near Christchurch in Canterbury, and also from Farewell Spit in north-west Nelson. The species was not seen in the South Island again until it was rediscovered on Farewell Spit in 1969.

Habitat: Strictly a coastal species, *Eleocharis neozelandica* inhabits the sandy margins of dune lakes and tidal creeks and damp sandy flats or hollows amongst sand dunes.

Conservation: Trampling of lake margins by livestock, raising and lowering of lake levels, and competition with aggressive weeds (such as Kikuyu grass) has probably contributed to the decline of this species. North Island population grows within a proposed scenic reserve, and a population near North Cape grows in a recreation area, but is threatened by trampling (by cattle and horses) and invasion by tree lupin. No specimens are known to be in cultivation.

Comments: *Eleo*: of swamps and marshes; *charis*: graceful; *neozelandica*: of New Zealand.

Status: Endangered.

M. Bellingham

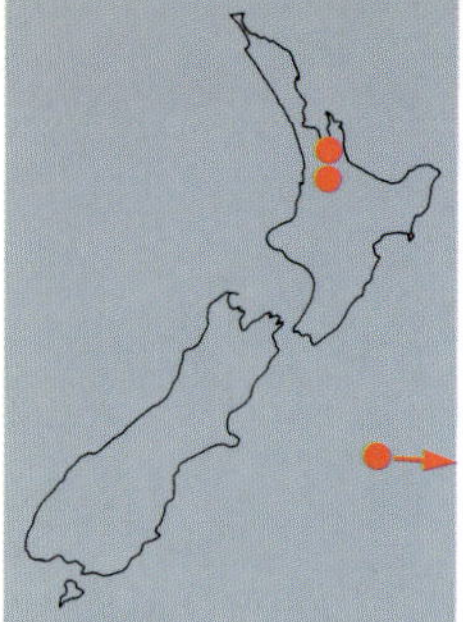

Sporadanthus traversii

Restionaceae

Description: Extremely robust, rush-like herb up to 3 m, erect, branching stems up to 10 mm in diameter. Stout, branching rhizomes form a "scaffolding" or "raft" held up from below by vertical stilt-like roots. Leaves reduced to tapering pointed sheaths up to 4 cm long, pressed closely against the stems, margins of the lowermost sheaths split to form a fringe. Tiny, inconspicuous flowers, each enclosed by a rigid bract, are in panicles at the tips of the stems; male and female flowers on separate plants. Fruit is a tiny capsule about 1 mm long and slightly less than 1 mm wide.
Flowers: October-December; Fruit: February-March.

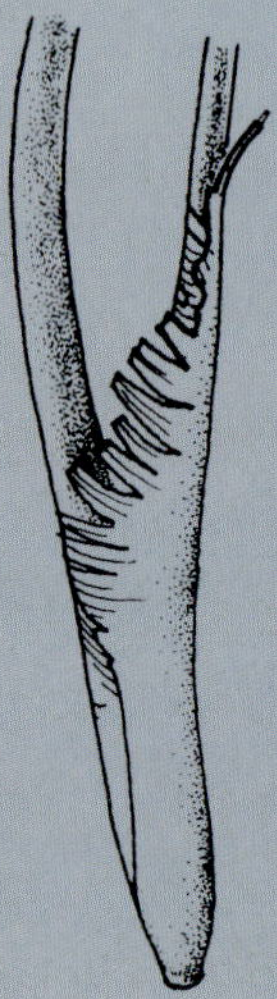

leaf sheath

E. Cameron P. Brooke

Distribution: *North Island, Chatham Islands. Endemic*. Bogs dominated by *Sporadanthus traversii* were formerly found near Kaitaia, in the Waikato lowlands and on the Hauraki Plains of the North Island. Some populations still remain in the Waikato and Hauraki lowlands but the Kaitaia population is now extinct. The species is widely scattered throughout Chatham Island in the Chatham group.

Habitat: *Sporadanthus traversii* is an important member of the plant communities on deep moist peat bogs. The interwoven "rafts" formed by the rhizomes of *S. traversii* may provide support for shrubs and also retain moisture and maintain waterlogged conditions in the bog.

Conservation: Although still abundant on the Southern Tablelands of Chatham Island, destruction of its habitat by systematic draining of swamps in the North Island and burning and drying out of restiad bogs on the Chatham Islands has led to the general decline of *Sporadanthus traversii*. Possible peat mining on the Chatham Islands, could threaten survival there. Two scientific reserves have been set up in the Waikato Valley but these are gradually drying out, as draining of the surrounding farmland lowers the water table. Very few specimens of *S. traversii* are known to be in cultivation.

Comments: *Sporadanthus traversii* is one of the few restiads in the New Zealand Botanical Region and is a distinctive element of the New Zealand flora. It is the only species in the genus. Wire-rush, a related species, resembles stunted *S. traversii* but has narrower and more flaccid stems which are usually less than 2.5 mm in diameter; the margins of its lower-most leaf sheaths are entire. *Sporadanthus*: scattered flowers; *traversii*: after Henry Travers (1844-1928), who is well known for his work on the Chatham Island flora.

Status: Vulnerable.

ferns and fern-allies

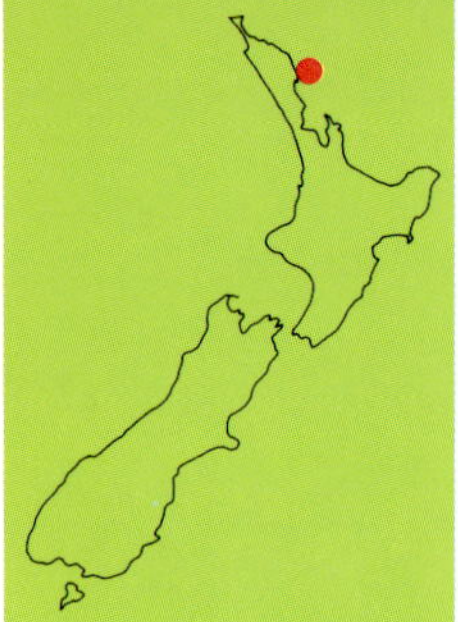

Asplenium pauperequitum

Aspleniaceae

Description: Tufts of glossy, dark green fronds arise from a short, erect rhizome bearing tapering, reddish brown scales up to 10 x 0.5 mm. Stiff, shiny, dark reddish brown stipes are 1-16 cm x 1 mm in diameter, scaly near the base but almost smooth further up and terminate in triangular laminae which are 3-10 x 2.5-6 cm. Primary pinnae are near the tip of each lamina in 1-5 approximately opposite pairs plus a large terminal lobe, each pinna elliptic or broadly ovate, margins entire or shallowly toothed. Lower primary pinnae are broadly ovate or triangular and may be 3-lobed or divided into a pair of shortly stalked secondary pinnae and a larger terminal lobe. Conspicuous sori, 4-12 x 1 mm, curve away from the midrib of each pinna.

Distribution: *Poor Knights Islands. Endemic. Asplenium pauperequitum* has been found only on the islands of Tawhiti Rahi and Aorangi in the Poor Knights group.

Habitat: This fern grows in crevices on exposed, southern-facing rhyolitic rock bluffs or in damp soil in rock crevices beneath coastal pohutukawa forest.

Conservation: Populations of this species are generally very small, but are fortunately afforded some protection by the nature reserve status of the Poor Knights Islands.

Comments: *Asplenium pauperequitum* was discovered in 1982 and described in the *New Zealand journal of botany* in 1984. *Asplenium*: against (complaints of) the spleen, referring to supposed medicinal properties; *pauperequitum*: (from the) Poor Knights (Islands).

Status: Vulnerable.

A. Wright

Cyathea kermadecensis

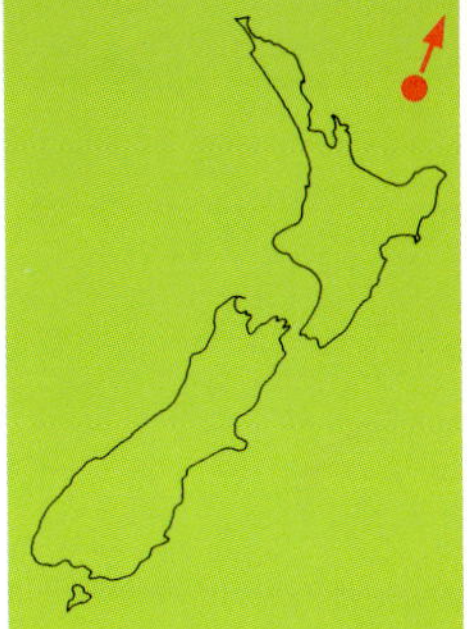

Cyatheaceae

Description: Slender tree fern with a trunk up to 20 m tall, surrounded at the base by a mass of rootlets. Deeply grooved stipes clad with numerous linear brown scales at the base, and higher up with smaller scales and woolly, yellowish brown hairs. Laminae rather tough, dull green on the upper surface, paler below, up to 4 m x 90 cm. Primary pinnae oblong-lanceolate, up to 45 x 14 cm. Secondary pinnae linear-oblong, slightly curved, up to 90 x 20 mm. They in turn are divided into curved, oblong tertiary pinnae up to 9 x 3 mm. The numerous sori are covered by membranous, cup-shaped indusia about 0.6 mm in diameter.

Distribution: *Kermadec Islands. Endemic.* This tree fern grows only on Raoul Island in the Kermadec group.

Habitat: *Cyathea kermadecensis* grows in wet forest and seems to be confined to a few high ridges and east-facing gullies.

Conservation: *Cyathea kermadecensis* once appeared to be abundant in the wetter parts of Raoul, but by 1966 it had become uncommon or absent in many areas as a result of browsing by goats. Probably less than 50 plants now remain in 2 major areas, and there is very little regeneration. The Kermadec Islands are a nature reserve and the last goat on Raoul Island was eliminated in November of 1984.

Comments: This species is very similar to *Cyathea milnei* which is also endemic to Raoul Island. *C. milnei*, however, has more gradually tapering secondary pinnae and high, almost globose indusia which shed their spores through irregular apertures with jagged margins. The indusia found on fronds of *C. kermadecensis* are slightly flattened and split along a regular line of dehiscence to give two smooth margins to the aperture. *C. kermadecensis* is closely related to *C. cunninghamii* which occurs in Australia and the North and South Islands of New Zealand but is not found on the Kermadec Islands. *Cyathea*: cup-like, referring to the shape of the indusia; *kermadecensis*: from the Kermadec Islands.

Status: Endangered.

W. Sykes

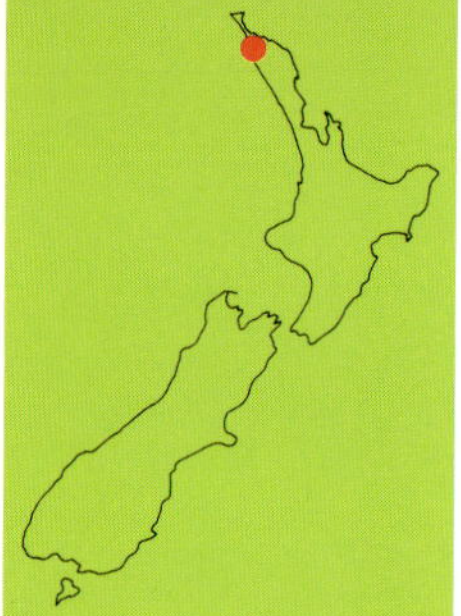

Christella dentata

soft fern

Thelypteridaceae

Description: Tufts of pale green, velvety textured fronds arising from a very shortly creeping rhizome. Laminae, supported on slender, greenish stipes 10-60 cm long, are oblong-lanceolate, to 1 m x 30 cm, rachis greenish and usually covered in slender hairs. Primary pinnae up to 15 x 2 cm, lower-most 2-4 pairs slightly smaller than the rest, primary pinna is divided into slightly curved, oblong secondary pinnae usually less than 1 cm long. Up to 7 pairs of sori are produced on each secondary pinna about half-way between the margin and the midrib. The sori are covered by heart- to kidney-shaped indusia and do not coalesce.

Distribution: *North Island. Abundant in tropical and semi-tropical regions*. Soft fern was found at several localities near Kaitaia, North Auckland, in the early 1900s, but in recent years it has been recorded from only three sites in this region.

Habitat: The only wild populations known in New Zealand grow in disturbed remnant kahikatea stands, sheltered in holes left by the rotting stumps of felled kahikatea.

Conservation: Less than 50 plants are estimated to exist in the wild in New Zealand. Two colonies have been invaded by grasses and other exotics and are heavily grazed by cattle. A few specimens are in cultivation.

Comments: This tropical species is closely related to another form of *Christella* which is also placed in *C. dentata* at present. This other *Christella* is found on geothermal sites in the central North Island of New Zealand, and probably occurs on Norfolk Island and in forest clearings on the Kermadec Islands. It differs in having fronds which are distinctly narrowed towards the base, often darker green, and an erect rhizome. Until recently, *Christella dentata* was included in the genus *Thelypteris*. *Christella*: after Hermann Christ (1833-1933), a Swiss pteridologist and taxonomist; *dentata*: toothed.

Status: Endangered.

D. Given

J. Miles

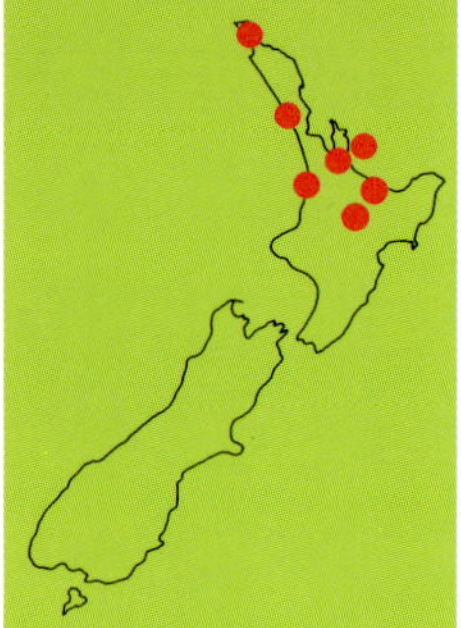

Cyclosorus interruptus

Thelypteridaceae

Description: Attractive fern forming spreading patches of rather stiff, erect, olive-green fronds up to 80 cm long and arising from a stout creeping rhizome. Fronds have slender stipes up to 60 cm x 5 mm, almost black at the base and brownish above. Supported on each stipe is a distinctive wedge-shaped, triangular to oblong lamina, 12-45 x 5-24 cm with 10-15 pairs of tough, narrowly lanceolate-shaped primary pinnae. Primary pinnae up to 15 x 1.5 cm. Each primary pinna divided into triangular to oblong secondary pinnae about 5 mm long. Numerous close-set sori, less than 1 mm in diameter, nearer the midrib than the margin, sometimes coalescing. The indusium covering each sorus is heart- or kidney-shaped.

Distribution: *North Island. Abundant in most tropical and warm temperate countries*. Colonies of this fern are scattered between Cape Reinga and Taupo in the northern half of the North Island, and it also grows on Mayor Island.

Habitat: In the central North Island *Cyclosorus interruptus* grows in swamps where the water has been heated by geothermal activity. It has been found in heated mud, warm water, and at sites where the fronds are constantly enveloped in steam. Further north the species grows in the absence of thermal activity, but only in swamps near the coast where there are no frosts.

D. Given

Conservation: Many of the colonies which are known today are small, but some large colonies of up to several hundred plants have been found. In 1977 a large population at Taupo was almost destroyed when a popular tourist area was landscaped and a small colony near Spirits Bay was destroyed by fire and draining in connection with pine planting in the 1970s. Many of the sites where *Cyclosorus interruptus* occurs are privately owned, but one large population is within a scenic reserve and another is within a scientific reserve. Several populations occur in state forests and wildlife management reserves. *C. interruptus* can be readily grown in a container or artificial swamp and is occasionally cultivated.

Comments: Until recently *C. interruptus* was included in the genus *Thelypteris*. It is found in some texts under the name *Thelypteris (*or *Cyclosorus) gongylodes*. *Cyclosorus*: a round sorus; *interruptus*: interrupted or discontinuous possibly referring to the shape of the frond which is abruptly truncated at the base.

Status: Vulnerable.

Doodia aspera

Blechnaceae

Description: Harsh-textured, pale green and more or less erect fronds arise in tufts from a short rhizome densely covered in black scales and producing long stolons. Blackish brown stipes are up to 22 cm, and are covered in black or dark brown scales at the base and in rough tubercles higher up. Laminae are linear-lanceolate to elliptic, 12-40 cm long, with 17-32 pairs of pinnae. The middle and upper pinnae, the longest of which are up to 6 cm long and 0.6 cm wide, are linear, margins sharply toothed. Lowest pair of pinnae triangular or rounded with sharply toothed margins. All the pinnae, except sometimes the lower-most pair, lack stalks and have widely flanged bases. Oblong to kidney-shaped sori up to 2 mm long occur in 1-2 parallel rows on either side of the midrib of each pinna.

Distribution: *Possibly no wild populations remain in New Zealand. Found in eastern Australia and on Norfolk Island and Lord Howe Island.* Populations of *Doodia aspera* have been recorded near North Cape, Chelsea, and Kawakawa in North Auckland. The last remaining locality where this species was known to occur was near Kawakawa.

Habitat: In New Zealand *Doodia aspera* was found in dry manuka scrubland and in open areas.

Conservation: The sole remaining population may have been destroyed in the early 1980s when the site near Kawakawa was burnt over and planted with exotic pines. *Doodia aspera* is easy to grow but very few plants of New Zealand origin are in cultivation.

Comments: *Doodia*: after Samuel Doody (1656-1706), a London bookseller who was Curator of the Chelsea Botanic Gardens; *aspera*: rough, referring to the tubercles on the stipes which give them a rough texture.

Status: Endangered.

P. Brownsey

Lycopodium serpentinum

Lycopodiaceae

Description: Very small club moss forming patches of rigid, erect, and sparingly branched aerial stems up to 10 cm arising from a prostrate, creeping main stem. Leaves about 3 mm long, linear-lanceolate, tapering to a fine point. Leaf-like sporophylls which bear sac-like sporangia containing the spores are aggregated into erect strobili borne at tips of the aerial stems. The strobili are usually solitary and up to 4 cm long. Each sporophyll is broadly triangular, 2-3 mm long, and tapers to a long sharp point at the tip.

Distribution: *North Island. Also found in Australia and New Caledonia*. Less than 10 populations of *Lycopodium serpentinum* are known in the North Auckland region on Aupouri Peninsula, Karikari Peninsula, and near Kaitaia and Ahipara. Several sites occur in the Te Kauwhata-Hamilton-Te Awamutu and Hauraki Plains regions of South Auckland.

Habitat: This club moss grows on open sites in peat bogs amongst low-growing umbrella ferns, sundews, and other peatland plants.

Conservation: Several of the sites where *Lycopodium serpentinum* occurs are threatened by draining for farming or peat mining. Fortunately one of the largest populations in New Zealand is within a scientific reserve and a second sizeable colony is in a historic reserve. Another population growing on a Department of Lands and Survey farm development block has been fenced to exclude cattle. No specimens are known to be in cultivation.

D. Given

Comments: *Lycopodium serpentinum* is one of New Zealand's smallest club mosses. *Lycopodium*: wolf's foot, referring to the roots which resemble a wolf's paw; *serpentinum*: snake-like or creeping.

Status: Vulnerable.

Ophioglossum petiolatum

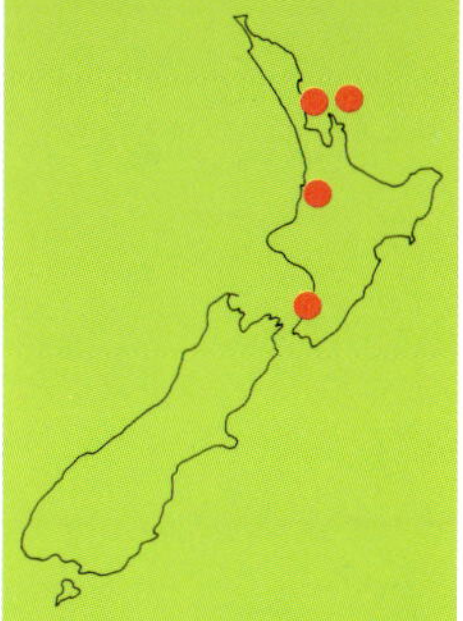

Ophioglossaceae

Description: An unusual herb-like fern up to 30 cm high with 1 or sometimes 2 leaves. The stem, which arises directly from the short cylindrical rhizome, is 1.5-16 cm long and divided near the top into 2 distinct portions; a fleshy, sterile lamina and a slender stalk which is 4-13.5 cm and terminates in a fertile spike. Lamina narrowly ovate, ovate or elliptic in outline, truncate at the base, 1.5-10.5 x 0.6-3.5 cm, longer laminae also having a short petiole up to 2.5 cm. Each fertile spike is 1.5-4 cm long and consists of 2 rows of 15-45 sporangia.

Distribution: *North Island. Also occurs in the Pacific Islands and throughout south-east Asia.* This species was once found at several sites in the North Island and was collected from at least three sites in the northern half of the South Island before 1950, but in the last few years it has only been found definitely on the Horowhenua coast in the south-west of the North Island, east of Kawhia Harbour, at a few sites near the northern end of Great Barrier Island, and along the coast of the Hauraki Gulf.

Habitat: *Ophioglossum petiolatum* has been recorded from moist talus and grassy areas, sandy margins of coastal lagoons, herbfields near lake margins, swamps and streams, and rarely podocarp forest.

Conservation: The 30-40 plants which have been found on the Horowhenua coast are threatened by the possible establishment of a pine plantation. Fortunately a small population near Kawhia Harbour grows within a scenic reserve. The plants on Great Barier Island seem relatively secure, as they occur in a rather remote area which will soon be given reserve status. However, these populations require monitoring to ensure that they are not overwhelmed by other plants or feral goats and pigs.

Comments: *Ophioglossum coriaceum* is closely related to *O. petiolatum*, but has smaller laminae, a shorter petiole, and less than 17 pairs of sporangia in the fertile spike. *O. petiolatum* is widespread in the tropics, but its exact distribution is difficult to determine, since it is often confused with *O. reticulatum*. The species is sometimes included in *O. pedunculosum*, in New Zealand floras. *Ophioglossum*; snake's tongue, referring to the appearance of the fertile spike; *petiolatum*; stalked, referring to the long petiole.

Status: Vulnerable

E. Cameron

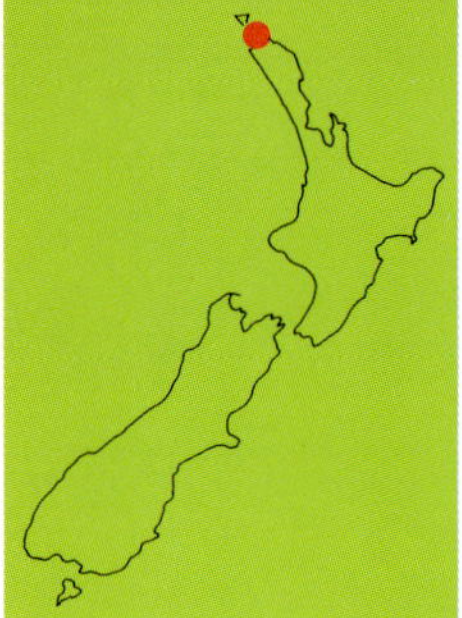

Phylloglossum drummondii

Lycopodiaceae

Description: Very small plant up to 7 cm high with small tubers and a rosette of up to 10 bright green, fleshy leaves. Leaf 1-2 cm long, linear, circular in cross section, and has a sharply pointed tip. A single leafless stalk up to 4 cm long arises from the centre of the rosette and terminates in an erect strobilus. The strobilus is 4-7 mm long, oblong-ovoid, and consists of up to 30 sharply pointed bracts which cover the sac-like sporangia containing the spores.

Phylloglossum drummondii has a very short growing season, only appearing above ground from May to October.

Distribution: *North Island. Also found in Australia and Tasmania*. This plant has been collected near Christchurch and in the Marlborough Sounds in the South Island and may once have been widespread throughout the North Island. Today it is restricted to the North Auckland region.

Habitat: Although *Phylloglossum drummondii* can survive in a fairly wide range of habitats, in North Auckland it is most commonly found in recently burnt-over scrubland, usually growing with sedges under sparse, low manuka less than 45 cm high, on seasonally damp gumland sites.

Conservation: This may be a difficult species to reserve without deliberate management to maintain a suitable habitat, since the plants tend to die out if they are shaded by taller vegetation. The virtual restriction of *Phylloglossum drummondii* to habitats dependent on fire could make it even rarer than it is at present, as more marginal scrubland is converted to farmland and as other areas, once protected, revert to tall scrub and forest. Substantial colonies occur in a historic reserve near Ahipara and in a proposed scientific reserve near Lake Ohia, on the Karikari Peninsula. The plants near Lake Ohia recently suffered severe browsing by insects or snails. Almost all strobili were destroyed. *P. drummondii* is difficult to cultivate; only a few specimens are in cultivation.

Comments: *Phylloglossum drummondii* is the only species in the genus. *Phylloglossum*: tongue-like leaf; *drummondii*: after James Drummond (1784-1863), a plant collector who was born in Scotland and emigrated to western Australia in 1829.

Status: Vulnerable.

J. Braggins

Todea barbara

king fern, hard todea

Osmundaceae

Description: Large fern up to 2 m tall forming clumps of dark green, somewhat leathery fronds arising from a stout, erect rhizome up to 120 cm tall and 60 cm wide. Stipes smooth, brown, up to 60 cm long and 1 cm or more in diameter. Primary pinnae 10-23 cm long, secondary pinnae bluntly toothed, linear-lanceolate, 1.5-5 cm x 2-6 mm wide. Sporangia crowded along veinlets of the lowermost secondary pinnae, often covering the whole lower surface of each pinna.

Distribution: *North Island, Poor Knights Islands. Also found in Australia, Tasmania, and the old world tropics.* Scattered populations of king fern occur between North Cape and the Bay of Islands in North Auckland and another colony grows on Aorangi Island in the Poor Knights group.

Habitat: In the far north *Todea* grows in drier sites amongst gumland scrub. It is most commonly found on the sides of gullies and old gumdigger's drains.

Conservation: Once found in numerous localities in the far north of the North Island, *Todea barbara* is now definitely known from only about 20 sites and its range is expected to decline still further as more land is cleared for farming. A good colony once grew near Coopers Beach but the plants were eliminated during the construction of a parking area and lookout in the late 1970s. In 1983 another large colony was destroyed when shrubland near Te Kao was cleared for planting in pines. One population near the Bay of Islands has been reserved in a forest sanctuary, a second population grows in a proposed scientific reserve, another occurs in a state forest, and a fourth is within a coastal farm park. The Poor Knights Islands are a nature reserve. *Todea barbara* is in cultivation. The species is readily raised from fresh spores and is not difficult to grow providing it is sheltered from frosts.

Comments: *Todea*: after Heinrich Tode (1733-1797), a German mycologist; *barbara*: foreign.

Status: Vulnerable.

J. Braggins

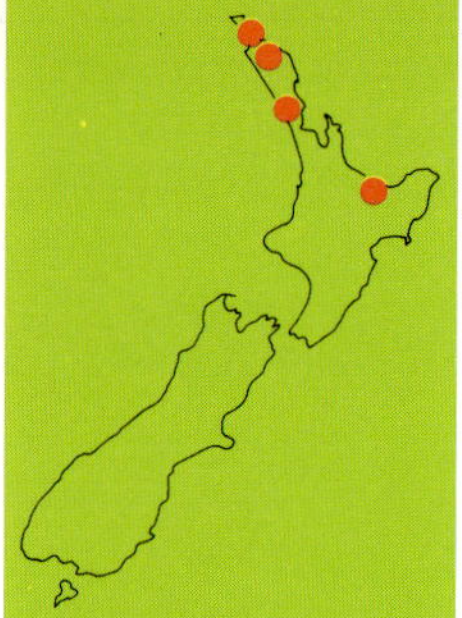

J. Miles

Thelypteris confluens

marsh fern

Thelypteridaceae

Description: Fern forming open patches of thin, softly-textured, very pale green fronds arising from slender, branched rhizomes which creep along the ground for considerable distances. Frond tips slender, pale brown, hairless up to 60 cm long, lamina oblong-lanceolate, 10-40 x 5-15 cm wide. Lanceolate-oblong primary pinnae, spread almost at right angles to the midrib of each frond, 2-8 cm x 5-15 mm wide, reduced in size at the base of the lamina. They are divided into linear-oblong secondary pinnae up to 1 cm long and usually with partially recurved margins. Close-set sori are barely 1 mm in diameter and often coalesce to form compound sori up to 5 mm long. The indusium covering each sorus is heart- or kidney-shaped.

Distribution: *North Island. Also found in Australia and South Africa*. Marsh fern may once have been moderately common in lowland swamps of the North Island to at least as far south as the Bay of Plenty and Wanganui. In 1872, for instance, Thomas Kirk recorded it in very large numbers over hundreds of hectares of coastal swampland near Maketu. These wetlands and many others gradually disappeared and by the 1930s *Thelypteris confluens* was a rare plant. Remnant populations of *T. confluens*, some of which are large, are scattered throughout coastal areas of North Auckland and also occur near Kaikohe. Several other colonies have been found further south in the Bay of Plenty and Rotorua regions.

Habitat: As its name implies, marsh fern grows in lowland swamps.

Conservation: Several populations of this fern are in state forests or on land recommended for reserve status. A few plants are within a wildlife management reserve and numerous plants are within Te Paki Farm Park, but many of the plants at a further site have been destroyed by cattle trampling. A few specimens are in cultivation.

Comments: Marsh fern is closely related to *Thelypteris palustris* of the Northern Hemisphere and has appeared under this name in older texts, sometimes as var. *squamigera*. *Thelypteris*: lady fern; *confluens*: confluent, referring to the coalescent sori.

Status: Vulnerable.

D. Given

Cyathea kermadecensis. *(R. Lamberts)*

Asplenium pauperequitum. *(R. Lamberts)*

Christella dentata.
(R. Lamberts)

Cyclosorus interruptus. *(R. Lamberts)*

Sori of illustrated ferns.

Doodia aspera. *(R. Lamberts)*

Ophioglossum petiolatum. *(P. Brownsey)*

Thelypteris confluens. *(R. Lamberts)*

Todea barbara. *(J. Braggins)*

Sori of illustrated ferns.

parasites and semi-parasites

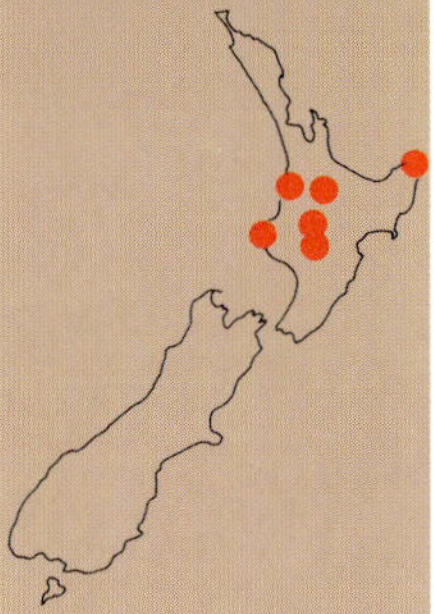

Dactylanthus taylorii

pua reinga, pua o te reinga; flower of Hades, wood rose

Balanophoraceae

Description: A root parasite which forms a warty, dark brown, ball-like structure up to 30 cm in diameter just below the surface of the ground and firmly attached to the expanded end of a root of the host tree. Fleshy, unbranched shoots which emerge between the warts of the surface are up to 20 cm long and covered in overlapping, pinkish brown, scale-like leaves up to 15 mm long. When these shoots break through the surface of the ground the uppermost leaves part to expose spikes of many tiny flowers. Male flowers produce mealy white pollen and may sometimes be detected by their characteristic fragrance. Dull reddish female flowers, produced on separate plants, are followed by ovoid, reddish brown nuts up to 2 mm long.
Flowers: January-April; Fruit: February-May.

Distribution: *North Island. Endemic.* Although it has been recorded in the past from Hokianga Harbour south to Mount Egmont and eastwards to near East Cape, its present distribution seems to be centred on Tongariro National Park, the ranges west of Lake Taupo, the Kaimanawa Range, and the upper Wanganui River. Outlying populations occur in the Kawhia and East Cape areas and on Mount Egmont.

Habitat: *Dactylanthus taylorii* is most commonly found in second growth forest and along the outer fringes of montane forest. It is usually parasitic on lancewood, five-finger, mountain five-finger, pate, mapou, *Pittosporum* spp., and *Coprosma* spp., but other hosts include hangehange, broadleaf, and *Nothofagus* spp.

Conservation: Widespread clearance of the North Island forests for milling and to provide farmland has eliminated this species in many areas. Forests are still being cleared or modified in some areas within the range of *Dactylanthus taylorii.* Another serious threat to the species survival is depletion of colonies by collectors of "wood roses" (obtained by boiling the parasite off the host root to leave a fluted, rose-like disc). Populations of *D. taylorii* are currently known from Tongariro National Park, Egmont National Park, and a few state forests and scenic reserves. Others probably still occur in Urewera National Park and Coromandel Forest Park even though it has not been reported from these areas recently. No specimens are known to be in cultivation.

Comments: *Dactylanthus taylorii* is New Zealand's only representative of the Balanophoraceae, a tropical family of parasitic plants. *Dactylanthus*: finger flower; *taylorii*: after Richard Taylor (1805-1873), a "missionary tramper" who visited New Zealand three times between 1839 and 1873.

Status: Vulnerable.

J. Bedford

flowers

J. Bedford

plant

J. Jansen

"wood rose"

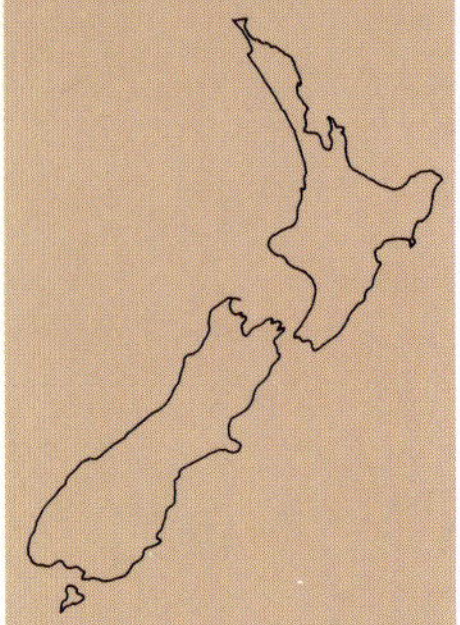

Trilepidea adamsii

Loranthaceae

Description: Bushy, semi-parasitic shrub, up to 1 m in diameter, branchlets circular in cross section. Leaves have stout, winged petioles up to 5 mm long, and broadly elliptic or diamond-shaped laminae 3.5-8 x 1-4 cm, margins usually recurved. Clusters of 2-4 yellowish flowers, tinged with reddish green, are produced in the leaf axils. Flowers 3-4 cm long, tubular near the base, slightly swollen in the middle, 4 slender recurved lobes above. Three small bracts surround the flower base. Drupe-like fruit, ellipsoid, 8-9 mm long, and red. Flowers: September-November; Fruiting period not known.

Distribution: *No wild populations known. Endemic.* This species was first collected near Thames in the late 1800s and was later reported from Waipoua and Kaipara in North Auckland, Coromandel Peninsula, Hunua and Maungakawa in South Auckland, and also from Great Barrier and Waiheke Islands. It has not been seen since 1954.

Habitat: *Trilepidea adamsii* was usually recorded as being semi-parasitic on mamangi, *Melicope* spp., and mapou.

Conservation: This species may have always been somewhat rare and local in distribution. It was probably depleted by the destruction of its habitat as forest was cleared for milling, farming, and gold mining, and through browsing by the Australian brush-tailed possum and over-collecting. No specimens are known in cultivation.

Comments: The genus is monotypic. *Trilepidea adamsii* has in the past been included in the genus *Elytranthe. Trilepidea*: with three scales, referring to the three bracts at the base of each flower; *adamsii*: after James Adams (1839-1906), a well-known New Zealand botanist who was Headmaster of the Church of England Grammar School in Parnell and Thames High School.

Status: Presumed extinct.

From Hetley, "The Native Flowers of New Zealand".

Botanical names of species referred to in the text by a common name only

Common name	Botanical name	Other common and Maori names
*black beech	_Nothofagus solandri_ var. _solandri_	tawhairauriki
blackberry	*_Rubus_ species	
bog pine	_Halocarpus bidwillii_ (formerly _Dacrydium bidwillii_)	
bracken	_Pteridium esculentum_	rahurahu
broadleaf	_Griselinia littoralis_	kapuka papauma
broom	*_Cytisus scoparius_	
five-finger	_Pseudopanax arboreus_	whauwhaupaku
gorse	*_Ulex europaeus_	
hangehange	_Geniostoma ligustrifolium_	New Zealand privet
kahikatea	_Dacrycarpus dacrydioides_ (formerly _Podocarpus dacrydioides_)	white pine
kanuka	_Kunzea ericoides_ (formerly _Leptospermum ericoides_)	kopuka white tea tree
karaka	_Corynocarpus laevigatus_	
kauri	_Agathis australis_	koroi
Kikuyu grass	*_Pennisetum clandestinum_	
kowhai	_Sophora_ species	
kumarahou	_Pomaderris kumeraho_	gum digger's soap
lancewood	_Pseudopanax crassifolius_	horoeka
mairehau	_Phebalium nudum_	
mamangi	_Coprosma arborea_	
manuka	_Leptospermum scoparium_	red tea tree
mapou	_Myrsine australis_	mapau matipou mataira takapou tipau
matagouri	_Discaria toumatou_	tumatakuru tuturi wild Irishman
mingimingi	_Leucopogon fasciculatus_ (formerly _Cyathodes fasciculata_)	
mist flower	*_Eupatorium riparium_	
mountain five-finger	_Pseudopanax colensoi_	orihou
oxeye daisy	*_Leucanthemum vulgare_ (formerly _Chrysanthemum leucanthemum_)	
pate	_Schefflera digitata_	patate kotete seven-finger
pigeonwood	_Hedycarya arborea_	porokaiwhiri
pohutukawa	_Metrosideros excelsa_	New Zealand Christmas tree
puriri	_Vitex lucens_	kauere
sea rocket	*_Cakile edentula_	
taraire	_Beilschmiedia tarairi_	
tawa	_Beilschmiedia tawa_	
toatoa	_Phyllocladus glaucus_	
tree lupin	*_Lupinus arboreus_	
wire rush	_Empodisma minus_ (formerly _Calorophus minor_)	

* = adventive species

Glossary

achene - a dry, single-seeded fruit that does not split open.
adventive - plant which grows in the wild but whose presence is directly or indirectly attributable to man.
alluvial soils - material deposited by rivers and streams.
alternate - arranged singly along a stem, not in opposite pairs or whorls.
anther - the terminal, pollen-bearing part of a stamen.
apex - the tip or free end of an organ.
asexual reproduction - any reproductive process that does not involve the union of gametes or sex cells.
awn - a stiff or bristle-like projection, often from the tip or back of an organ.
axis - the main stem or central longitudinal support.

beak - a pointed projection or prolongation of an organ.
bract - a modified, often much reduced leaf.

calyx - the outer whorl of flower parts. Made up of sepals and usually greenish.
capitulum - a "flower-like" aggregation of small simple flowers forming a compact head (Asteraceae).
carpel - one of the flower's female reproductive organs, consisting of an ovary containing one or more ovules, a stigma, and often a stalk-like style between them.
capsule - a dry fruit which splits open to shed the seeds.
caryopsis - an achene which is closely fused to the thin wall of the fruit; characteristic of the grasses (Poaceae).
coccus - one of the parts of a dry, lobed fruit, each part single-seeded. Plural: cocci.
colluvial soils - material at the foot of a cliff or steep slope that has been transported by gravity.
column - a structure within an orchid flower, formed by the union of stamens, styles, and stigmas.
compound leaf - a leaf divided into several leaflets.
conglomerate - rounded shingle which has been compressed over a long period to form compact rock.

declinate - curved or bent downwards.
decumbent - prostrate stems or branches with the tips turning upwards from the soil.
dehiscent - opening or splitting into separate parts to shed contents; hence dehiscence.
dicotyledon - a plant whose embryo has two "seed leaves".
disc-florets - the inner flowers of the flower head in many members of the daisy family (Asteraceae), as distinct from the surrounding petal-like ray-florets.
divaricating - intertangled stems spreading at a very wide angle.
drupe - a "stone-fruit" with a single seed encased in a hard coat and surrounded by a fleshy layer or leathery skin, e.g., a plum.

embryo - the rudimentary plant within the seed.
endemic - native and restricted to a particular county or region; in this book meaning occurring naturally only in the New Zealand Botanical Region or some part of it.
endocarp - the innermost layer of the fruit wall which lies next to the seed and is sometimes hard and "stony".
entire - having a continuous margin, without teeth or divisions of any kind.
epiphyte - a plant that grows on the surface of another, without deriving food from living tissues of the host.

family - a category in plant classification above the level of genus, a family contains one to many genera and has a unique combination of characteristics which distinguishes it from all other families.
filament - the anther-bearing stalk of a stamen.
flaccid - limp, not rigid.
flexuous - having a wavy or zig-zag form.
floret - an individual flower in a flower head (e.g., Asteraceae) or grass spikelet (Poaceae).
flower head - an inflorescence of small flowers densely clustered on the top of the receptacle, as in the daisy family (Asteraceae).
follicle - a dry dehiscent fruit which is derived from one carpel and which splits open along one side.
frond - a leaf of a fern.

genus - a category in plant classification above the level of species and below that of family; a genus contains one to many species and has a unique combination of characteristics which distinguishes it from every other genus. Plural: genera.
glume - one of two small scales or bracts at the base of the whole spikelet in the grasses (Poaceae), or a single bract subtending each floret in the sedges (Cyperaceae).
gumland - land which was once covered with kauri forest and now contains deposits of fossilised kauri gum.

habit - the general form and shape of a plant.
habitat - the type of place in which a plant grows.
herb - any higher plant whose stems do not become hard and woody; hence herbaceous.
herbfield - vegetation distinguished by the abundance of various large herbs.
hybrid - a plant whose parents are of different taxa.

indusium - a flap or cup-shaped structure more or less covering each sorus in some ferns. Plural: indusia.
inflorescence - a general term for the arrangement of the flowers.
internode - the part of a stem between two nodes.

labellum - the modified middle or lower petal of an orchid flower which usually differs in size and shape from the lateral petals.
lamina - the flat, expanded portion or blade of a leaf or frond. Plural: laminae (see p.146).
layer - a woody stem which takes root while still attached to the parent plant. The process is known as layering.
leaf axil - the upper angle between a leaf petiole and the stem to which it is attached.
leaflet - one leaf-like portion of a compound leaf.
leaf sheath - the basal portion of a leaf blade or petiole; it more or less completely surrounds the stem.
lemma - the lower and larger of two scales or bracts enclosing each floret in a grass spikelet (the whole spikelet is usually subtended by two glumes).
liane - a woody climbing plant.
lobe - a division of a leaf or leaf part, especially when rounded.
locule - an enclosed compartment within an organ, as in an ovary or fruit.

mericarp - a single-seeded portion of a schizocarp.
midrib - the main or central vein of a leaf or pinna, etc.
monocotyledon - a plant whose embryo has only one "seed leaf".

native - where presence cannot be attributed to man.
naturalised - brought into a region by man but now well established in the wild.
New Zealand Botanical Region - for the purposes of this book this consists of the Kermadec and Three Kings Islands, the North and South Islands and adjacent islands, Stewart Island, the Chatham Islands, and the Subantarctic Islands, including the Snares, Bounty, Antipodes, Auckland, and Campbell Islands. Macquarie Island is usually included in the New Zealand Botanical Region, but its plants have not been dealt with in this book, as the management and conservation of the island's flora comes under the jurisdiction of the Australian Government.
node - a place on a stem from which one or more leaves or bracts arise.
nut - a dry fruit with a hard, woody wall which does not split to release the single seed within. Hence nutlet: a small nut.

oblique - longer on one side than on the other, i.e., asymmetrical.
order - a category in plant classification above the level of family and below that of class; an order contains one to many families and has a unique combination of characteristics which distinguishes it from all other orders.
ovary - the hollow region containing the ovules, found at the base of the female part of a flower. After fertilisation it develops into a fruit containing seeds.
ovule - the reproductive structure which contains the egg and develops into a seed after fertilisation.

panicle - a branched inflorescence.
papilla - a minute pimple-like projection. Plural: papillae.
parasite - a plant that obtains all its food from another living plant to which it is attached (see semi-parasite).
pedicel - the stem of a single flower.

perennial - with a life-span of over 2 years.
perianth - the calyx and petals of a flower when referred to collectively; used especially when the calyx and petals are similar, or when either one is absent.
petal - one separate part of an inner whorl of flower parts, sometimes coloured or patterned.
petiole - the stalk of a leaf.
pinna - a subdivision, or leaflet, of a pinnate leaf or frond. Plural: pinnae.
pinnate - a compound leaf, with the parts (pinnae) arranged along both sides of an axis.
podocarp - species of plants, mostly trees, belonging to the family Podocarpaceae, which includes New Zealand native confers with more or less fleshy fruits.
prostrate - lying flat on the ground.

ray-floret - the outer, petal-like flowers of the flower head in members of the daisy family (Asteraceae).
receptacle - the expanded tip of the stem on which the flower or flower head is borne.
recurved - curved or bent backwards.
reflexed - bent sharply backwards.
restiads - members of the Restionaceae, a small family of rush-like herbs.
rhachis - the midrib of a fern frond, continuous with the stipe.
rhizome - an underground stem which usually grows more or less horizontally.
rhyolite - a volcanic rock with high silica and feldspar content.
rosette - a circle of leaves radiating out from a centre, usually at or near ground level.

scabrid - rough to the touch because of minute, stiff projections.
scale - a small, leaf-like organ, often dry and membranous.
schizocarp - a dry fruit which splits into single-seeded mericarps when ripe.
self-fertile - male and female gametes (contained in the pollen and ovules respectively) are able to fuse and eventually produce viable seed, even when both originate in the same plant.
semi-parasite - a plant that produces part of its food by photosynthesis and obtains the rest from another living plant to which it is attached, e.g., *Trilepidea adamsii*, a native mistletoe.
sepal - one separate part of the outer whorl of flower parts; usually green and leaf-like.
sessile - without a stalk or petiole.
sexual reproduction - reproduction involving the union of male and female gametes from two different parents. (Male gametes are contained in the pollen and female gametes are contained in the ovules.)
sheath - see leaf sheath.
silicle - a dry fruit almost as broad, or broader than it is long. Its two valves separate from the persistent central partition, e.g., species of *Lepidium*.
sinus - in *Hebe* leaf buds, the gap between the bases of the two leaves of a pair.
sorus - a cluster of sporangia, as in most ferns. Plural: sori.
species (sp.) - the basic unit of plant classification; a group of similar individuals or populations which are sufficiently closely related to be all potentially capable of interbreeding, but which breed less successfully with other such groups.
spike - an unbranched, elongated shoot of sessile flowers, e.g., *Plantago*. Spikes are also made up of sporangia in the Ophioglossaceae and spikelets in the grass family (Poaceae).
spikelet - a tiny spike; in the grasses (Poaceae) and sedges (Cyperaceae, but excluding the genera *Carex* and *Uncinia*), the flowers of each spikelet are very small and enclosed by overlapping scales (known as lemmas in the grasses and glumes in the sedges).
spore - an asexual, usually single-celled reproductive body.
sporangium - a tiny capsule containing spores. Plural: sporangia.
sporophyll - a modified, leaf-like structure bearing sporangia.
stamen - the male part of a flower which produces pollen and consists of an anther supported on a slender stalk or filament.
stigma - the part of a female organ in a flower which is receptive to pollen, usually found at the tip of the style.
stipe - the stalk of a fern frond; analogous to the petiole of a leaf.
stolon - a horizontal stem which runs along the ground surface and produces roots at its nodes.
strobilus - a shortened section of a stem bearing sporophylls. Plural: strobili.
style - part of the female reproductive organ of a flower which is situated above the ovary and bears the stigma.
sub- - prefix meaning somewhat, slightly, or not quite.
subspecies (subsp.) - a markedly distinct variant of a species, resulting from some barrier to successful breeding with other populations of the species - usually geographical isolation.
subtend - below, but close to, another organ.

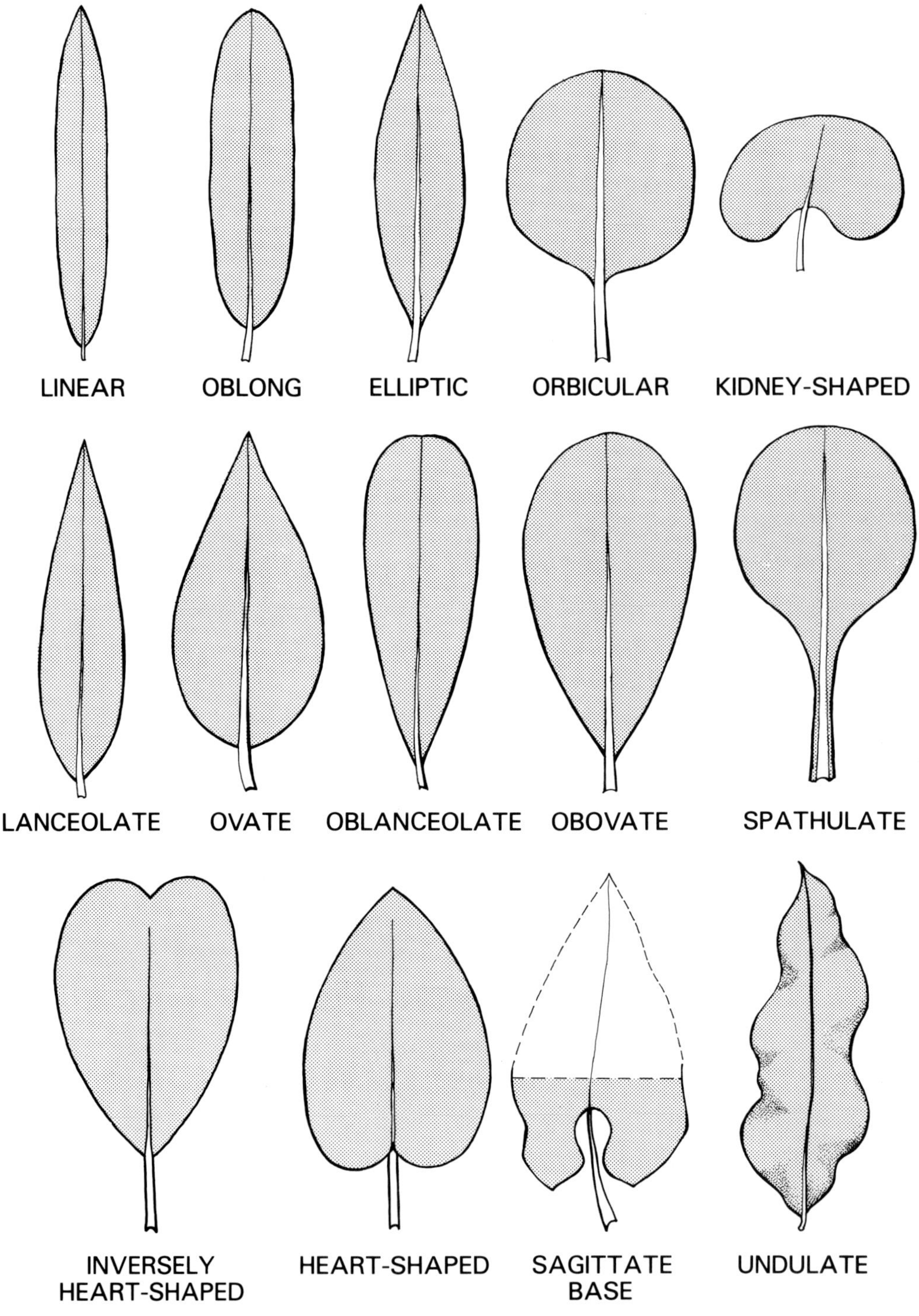

Diagrams explaining terms used to describe the shape of laminae. (P. Brooke)

succession - the sequence of changes in species dominance of a plant community during the development of vegetation in any area, including initial colonisation of a previously unoccupied area through to maturation of that vegetation.

talus - rock fragments of any size and shape (but usually coarse and angular) lying below the cliff or steep rocky slope from which they were derived.
taxon - a unit of any rank in a classification system, e.g., species, variety, genus, family. Plural: taxa.
tea-tree scrub - a dense, more or less continuous cover of manuka and/or kanuka not more than 5-6 m high.
tiller - a basal side shoot, as is found in the grasses (Poaceae).
tuber - an underground stem that persists from season to season and which is swollen with food reserves.
tuff - a sediment composed of volcanic ash and tiny fragments of volcanic rock.
turf - an even, mat-like covering of very short plants.
type locality - place from where the original specimen, on which a specific name is based, was collected.

undulate - waved in a plane at right angles to the surface.
utricle - a thin covering enveloping each tiny fruit or female flower in species of *Carex* and *Uncinia* of the sedge family (Cyperaceae).

valve - parts into which the wall of a capsule may split to release its seeds.
variety (var.) - a distinct geographical variant of a species or subspecies.
vegetative reproduction - propagation by vegetative tissues and organs, e.g., stems, leaves, tubers, rhizomes, etc.
viable - able to survive and develop.

whipcord Hebe - a *Hebe* with small scale-like leaves pressed closely against its branchlets.
whorl - a ring of leaves, petals, etc., where three or more arise at the same place on an axis.

Selected references

Bramwell, D.; Hamann, O.; Heywood, V.; Synge, H. (Eds) 1987: Botanic gardens and the world conservation strategy. New York, Academic Press.

Ehrlich, P.; Ehrlich, A. 1981: Extinction. New York, Random House.

Given, D. R. 1981: Rare and endangered plants of New Zealand. Auckland, Reed.

__________ (Ed.) 1983: Conservation of plant species and habitats. Wellington, Nature Conservation Council.

Given, D. R.; Williams, P. A. 1984: Conservation of Chatham Island flora and vegetation. Christchurch, Botany Division, Department of Scientific and Industrial Research.

Hall, A. V. (Ed.) 1984: Conservation of threatened natural habitats. South African National Scientific Programmes Report No. 92.

Huxley, A. 1984: Green inheritance. Glasgow, Collins.

Koopowitz, H.; Kaye, H. 1984: Plant extinction: a global crisis. Lexington, MA, USA, Stone Wall Press.

Leigh, J.; Boden, R.; Briggs, J. 1984: Extinct and endangered plants of Australia. London, Macmillan Press.

Lucas, G.; Synge, H. 1978: The IUCN plant red data book. International Union for Conservation of Nature.

Nature Conservation Council 1979: A vanishing heritage. Wellington, Nature Conservation Council.

Norton, B. G. 1987: Why preserve natural variety? Princeton, USA, University Press of Princeton.

Soulé, M. E. (Ed.) 1986: Conservation biology. Sunderland, Mass., USA, Sinaeur.

Synge, H. (Ed.) 1981: The biological aspects of rare plant conservation. New York, Wiley.

Taylor, P. W. 1986: Respect for nature. Princeton, USA, University Press of Princeton.

Williams, G. R.; Given, D. R. (Compilers) 1981: The red data book of New Zealand. Wellington, Nature Conservation Council.

Wilson, E. O. (Ed.) 1988: Biodiversity. Washington, National Academy Press, National Academy of Sciences.

Sources of advice

The following authorities can provide advice about conservation of threatened plants, and should be informed about any rare plants discovered. As Regional Offices of the Department of Conservation may have changed since this book went to press, please check their addresses with Central Office, Wellington.

Botany Division, DSIR,
Private Bag, CHRISTCHURCH

Botany Division Regional Station, DSIR,
Private Bag, AUCKLAND

Botany Division Regional Station, DSIR,
C/- Forest Research Institute,
Private Bag, ROTORUA

Botany Division Regional Station, DSIR,
Private Bag, HAVELOCK NORTH

Botany Division Regional Station,
C/- Division of Land and Soil Science, DSIR,
Private Bag, LOWER HUTT

Botany Division Regional Station, DSIR,
Private Bag, NELSON

Botany Division Regional Station, DSIR,
Private Bag, DUNEDIN

Department of Conservation, Central Office,
PO Box 10 420, WELLINGTON

Department of Conservation Regional Office,
Private Bag, Newton Post Office, AUCKLAND

Department of Conservation Regional Office,
Private Bag, HAMILTON

Department of Conservation Regional Office,
PO Box 1146, ROTORUA

Department of Conservation Regional Office,
Private Bag, WANGANUI

Department of Conservation Regional Office,
Private Bag, NELSON

Department of Conservation Regional Office,
Private Bag, HOKITIKA

Department of Conservation Regional Office,
Private Bag, CHRISTCHURCH

Department of Conservation Regional Office,
PO Box 1615, DUNEDIN

Forest Research Institute,
Private Bag, ROTORUA

Maruia Society,
PO Box 756, NELSON

Nature Conservation Council,
PO Box 12 200, WELLINGTON

New Zealand Natural Heritage Foundation,
Massey University,
PALMERSTON NORTH

Royal Forest and Bird Protection Society
of New Zealand,
Head Office, PO Box 631, WELLINGTON

World Wildlife Fund-New Zealand (WWF-NZ),
PO Box 6237, WELLINGTON

General index